Lecture Notes in Statistics

Edited by J. Berger, S. Fienberg, J. Gani, K. Krickeberg, and B. Singer

48

G. Larry Bretthorst

Bayesian Spectrum Analysis and Parameter Estimation

Springer-Verlag Berlin Heidelberg GmbH

Author

G. Larry Bretthorst
Department of Chemistry, Campus Box 1134, Washington University
1 Brookins Drive, St. Louis, MO 63130, USA

Mathematics Subject Classification: 62F15, 62Hxx

ISBN 978-0-387-96871-1 ISBN 978-1-4684-9399-3 (eBook)
DOI 10.1007/978-1-4684-9399-3

Originally published by Springer-Verlag Berlin Heidelberg in 1988.

2847/3140-543210

To E. T. Jaynes

Preface

This work is essentially an extensive revision of my Ph.D. dissertation, [1]. It is primarily a research document on the application of probability theory to the parameter estimation problem. The people who will be interested in this material are physicists, economists, and engineers who have to deal with data on a daily basis; consequently, we have included a great deal of introductory and tutorial material. Any person with the equivalent of the mathematics background required for the graduate-level study of physics should be able to follow the material contained in this book, though not without effort.

From the time the dissertation was written until now (approximately one year) our understanding of the parameter estimation problem has changed extensively. We have tried to incorporate what we have learned into this book.

I am indebted to a number of people who have aided me in preparing this document: Dr. C. Ray Smith, Steve Finney, Juana Sunchez, Matthew Self, and Dr. Pat Gibbons who acted as readers and editors. In addition, I must extend my deepest thanks to Dr. Joseph Ackerman for his support during the time this manuscript was being prepared.

Last, I am especially indebted to Professor E. T. Jaynes for his assistance and guidance. Indeed it is my opinion that Dr. Jaynes should be a coauthor on this work, but when asked about this, his response has always been "Everybody knows that Ph.D. students have advisors." While his statement is true, it is essentially irrelevant; the amount of time and effort he has expended providing background material, interpretations, editing, and in places, writing this material cannot be overstated, and he deserves more credit for his effort than an "Acknowledgment."

St. Louis, Missouri, 1988 G. Larry Bretthorst

Contents

1 INTRODUCTION 1
1.1 Historical Perspective 5
1.2 Method of Calculation 8

2 SINGLE STATIONARY SINUSOID PLUS NOISE 13
2.1 The Model 13
2.2 The Likelihood Function 14
2.3 Elimination of Nuisance Parameters 18
2.4 Resolving Power 20
2.5 The Power Spectral Density $\hat{p}$ 25
2.6 Wolf's Relative Sunspot Numbers 27

3 THE GENERAL MODEL EQUATION PLUS NOISE 31
3.1 The Likelihood Function 31
3.2 The Orthonormal Model Equations 32
3.3 Elimination of the Nuisance Parameters 34
3.4 The Bessel Inequality 35
3.5 An Intuitive Picture 36
3.6 A Simple Diagnostic Test 38

4 ESTIMATING THE PARAMETERS 43
4.1 The Expected Amplitudes $\langle A_j \rangle$ 43
4.2 The Second Posterior Moments $\langle A_j A_k \rangle$ 45
4.3 The Estimated Noise Variance $\langle \sigma^2 \rangle$ 46
4.4 The Signal-To-Noise Ratio 47
4.5 Estimating the $\{\omega\}$ Parameters 48
4.6 The Power Spectral Density 51

5 MODEL SELECTION **55**

5.1 What About "Something Else?" 55

5.2 The Relative Probability of Model f_j 57

5.3 One More Parameter 63

5.4 What is a Good Model? 65

6 SPECTRAL ESTIMATION **69**

6.1 The Spectrum of a Single Frequency 70

6.1.1 The "Student t-Distribution" 70

6.1.2 Example – Single Harmonic Frequency 71

6.1.3 The Sampling Distribution of the Estimates 74

6.1.4 Violating the Assumptions – Robustness 74

6.1.5 Nonuniform Sampling 81

6.2 A Frequency with Lorentzian Decay 86

6.2.1 The "Student t-Distribution" 87

6.2.2 Accuracy Estimates 88

6.2.3 Example – One Frequency with Decay 90

6.3 Two Harmonic Frequencies 94

6.3.1 The "Student t-Distribution" 94

6.3.2 Accuracy Estimates 98

6.3.3 More Accuracy Estimates 101

6.3.4 The Power Spectral Density 103

6.3.5 Example – Two Harmonic Frequencies 105

6.4 Estimation of Multiple
Stationary Frequencies 108

6.5 The "Student t-Distribution" 109

6.5.1 Example – Multiple Stationary Frequencies 111

6.5.2 The Power Spectral Density 112

6.5.3 The Line Power Spectral Density 114

6.6 Multiple Nonstationary Frequency Estimation 115

7 APPLICATIONS **117**

7.1 NMR Time Series 117

7.2 Corn Crop Yields 134

7.3 Another NMR Example 144

7.4 Wolf's Relative Sunspot Numbers 148

7.4.1 Orthogonal Expansion of the Relative Sunspot Numbers 148
7.4.2 Harmonic Analysis of the Relative Sunspot Numbers 151
7.4.3 The Sunspot Numbers in Terms of Harmonically Related Frequencies 157
7.4.4 Chirp in the Sunspot Numbers 158
7.5 Multiple Measurements 161
7.5.1 The Averaging Rule 163
7.5.2 The Resolution Improvement 166
7.5.3 Signal Detection 167
7.5.4 The Distribution of the Sample Estimates 169
7.5.5 Example – Multiple Measurements 173

8 SUMMARY AND CONCLUSIONS 179
8.1 Summary 179
8.2 Conclusions 180

A Choosing a Prior Probability 183

B Improper Priors as Limits 189

C Removing Nuisance Parameters 193

D Uninformative Prior Probabilities 195

E Computing the "Student t-Distribution" 197

List of Figures

2.1 Wolf's Relative Sunspot Numbers 28

5.1 Choosing a Model 66

6.1 Single Frequency Estimation 72
6.2 The Distribution of the Sample Estimates 75
6.3 Periodic but Nonharmonic Time Signals 77
6.4 The Effect of Nonstationary, Nonwhite Noise 79
6.5 Why Aliases Exist 82
6.6 Why Aliases Go Away for Nonuniformly Sampled Data 84
6.7 Uniform Sampling Compared to Nonuniform Sampling 85
6.8 Single Frequency with Lorentzian Decay 91
6.9 Two Harmonic Frequencies – The Data 106
6.10 Posterior Probability density of Two Harmonic Frequencies 107
6.11 Multiple Harmonic Frequencies 113

7.1 Analyzing NMR Spectra 119
7.2 The Log_{10} Probability of One Frequency in Both Channels 121
7.3 The One-Frequency Model 123
7.4 The Two-Frequency Model 125
7.5 The Three-Frequency Model 126
7.6 The Four-Frequency Model 127
7.7 The Five-Frequency Model 129
7.8 The Six-Frequency Model 130
7.9 The Seven-Frequency Model 131
7.10 Comparison to an Absorption Spectrum 132
7.11 Corn Crop Yields for Three Selected States 136
7.12 The Joint Probability of a Frequency Plus a Trend 139
7.13 Probability of Two Frequencies After Trend Correction 143

7.14 A Second NMR Example - Decay Envelope Extraction 145
7.15 How Does an NMR Signal Decay? 147
7.16 The Probability of the Expansion Order 150
7.17 Adding a Constant to the Model 152
7.18 The Posterior Probability of Nine Frequencies 155
7.19 The Predicted Sunspot Series 156
7.20 Chirp in the Sunspot Numbers? 160
7.21 A Simple Diffraction Pattern 164
7.22 Log_{10} Probability of a Single Harmonic Frequency 165
7.23 Example – Multiple Measurements 171
7.24 The Distribution of Sample Estimates 174
7.25 Example - Diffraction Experiment 176
7.26 Example - Two Frequencies 177

Chapter 1

INTRODUCTION

Experiments are performed in three general steps: first, the experiment must be designed; second, the data must be gathered; and third, the data must be analyzed. These three steps are highly idealized, and no clear boundary exists between them. The problem of analyzing the data is one that should be faced early in the design phase. Gathering the data in such a way as to learn the most about a phenomenon is what doing an experiment is all about. It will do an experimenter little good to obtain a set of data that does not bear directly on the model, or hypotheses, to be tested.

In many experiments it is essential that one does the best possible job in analyzing the data. This could be true because no more data can be obtained, or one is trying to discover a very small effect. Furthermore, thanks to modern computers, sophisticated data analysis is far less costly than data acquisition, so there is no excuse for not doing the best job of analysis that one can.

The theory of optimum data analysis, which takes into account not only the raw data but also the prior knowledge that one has to supplement the data, has been in existence – at least, as a well-formulated program – since the time of Laplace. But the resulting Bayesian probability theory (i.e., the direct application of probability theory as a method of inference) using realistic models has been little applied to spectral estimation problems and in science in general. Consequently, even though probability theory is well understood, its application and the orders of magnitude improvement in parameter estimates that its application can bring, are not. We hope to show the advantage of using probability theory in this way by developing a little of it and applying the results to some real data from physics and economics.

The basic model we are considering is always: we have recorded a discrete data set $D = \{d_1, \cdots, d_N\}$, sampled from $y(t)$ at discrete times $\{t_1, \cdots, t_N\}$, with a model equation

$$d_i = y(t_i) = f(t_i) + e_i, \qquad (1 \leq i \leq N)$$

where $f(t_i)$ is the signal and e_i represents noise in the problem. *Different models correspond to different choices of the signal* $f(t)$. The most general model we will analyze will be of the form

$$f(t) = \sum_{j=1}^{m} B_j G_j(t, \{\omega\}).$$

The model functions, $G_i(t, \{\omega\})$, are functions of other parameters $\{\omega_1, \cdots, \omega_r\}$ which we label collectively $\{\omega\}$ (these parameters might be frequencies, chirp rates, decay rates, the time of some event, or any other quantities one could encounter).

We have not assumed the time intervals to be uniform, nor have we assumed the data to be drawn from some stationary Gaussian process. Indeed, in the most general formulation of the problem such considerations will be completely irrelevant. In the traditional way of thinking about this problem, one imagines that the data are one sample drawn from an infinite population of possible samples. One then uses probability only for the distribution of possible samples that could have been drawn – but were not. Instead, what we will do is to concentrate our attention on the actual data obtained, and use probability to make the "best" estimate of the parameters; i.e. the values that were realized when the data were taken.

We will concentrate on the $\{\omega\}$ parameters, and often consider the amplitudes $\{B\}$ as nuisance parameters. The basic question we would like to answer is: "What are the best estimates of the $\{\omega\}$ parameters one can make, independent of the amplitudes $\{B\}$ and independent of the noise variance?" We will solve this problem for the case where we have little prior information about the amplitudes $\{B\}$, the $\{\omega\}$ parameters, and the noise. Because we incorporate little prior information into the problem beyond the form of the model functions, the estimates of the amplitudes $\{B\}$ and the nonlinear $\{\omega\}$ parameters cannot differ greatly from the estimates one would obtain from least squares or maximum likelihood. However, using least squares or maximum likelihood would require us to estimate all parameters, interesting and non-interesting, simultaneously; thus one would have the computational problem of finding a global maximum in a space of high dimensionality.

By direct application of probability theory we will be able to remove the uninteresting parameters and see what the data have to tell us about the interesting ones, reducing the problem to one of low dimensionality, equal to the number of interesting parameters. In a typical "small" problem this might reduce the search dimensions from ten to two; in one "large" problem the reduction was from thousands to six or seven. This represents many orders of magnitude reduction in computation, the difference between what is feasible, and what is not.

Additionally, the direct application of probability theory also tells us the accuracy of our estimates, which direct least squares does not give at all, and which maximum likelihood gives us only by a different calculation (sampling distribution of the estimator) which can be more difficult than the high-dimensional search one – and even then refers only to an imaginary class of different data sets, not the specific one at hand.

In Chapter 2, we analyze a time series which contains a single stationary harmonic signal plus noise, because it contains most of the points of principle that must be faced in the more general problem. In particular we derive the probability that a signal of frequency ω is present, regardless of its amplitude, phase, and the variance of the noise. We then demonstrate that the estimates one obtains using probability theory are a full order of magnitude better than what one would obtain using the discrete Fourier transform as a frequency estimator. This is not magic; we are able to understand intuitively why it is true, and also to show that probability theory has built-in automatic safety devices that prevent it from giving overoptimistic accuracy claims. In addition, an example is given of numerical analysis of real data illustrating the calculation.

In Chapter 3, we discuss the types of model equations used, introduce the concept of an orthonormal model, and derive a transformation which will take any nonorthonormal model into an orthonormal one. Using these orthonormal models, we then remove the simplifying assumptions that were made in Chapter 2, generalize the analysis to arbitrary model equations, and discuss a number of surprising features to illustrate the power and generality of the method, including an intuitive picture of model fitting that allows one to understand which parameters probability theory will estimate and why, in simple terms.

In Chapter 4 we calculate a number of posterior expectation values including the first and second moments, define a power spectral density, and we devise a procedure for estimating the nonlinear $\{\omega\}$ parameters.

In Chapter 5 we turn our attention to the problem of selecting the "best" model of a process. Although this problem sounds very different from the parameter estimation problem, it is essentially the same calculation. Here, we compute the relative posterior probability of a model: this allows one to select the most probable model based on how well its parameters are estimated, and how well it fits the data.

In Chapter 6, we specialize the discussion to spectral estimates and, proceeding through stages, investigate the one-stationary-frequency problem and explicitly calculate the posterior probability of a simple harmonic frequency independent of its amplitude, phase and the variance of the noise, without the simplifying assumptions made in Chapter 2.

At that point we pause briefly to examine some of the assumptions made in the calculation and show that when these assumptions are violated by the data, the answers one obtains are still correct in a well-defined sense, but more conservative in the sense that the accuracy estimates are wider. We also compare uniform and nonuniform time sampling and demonstrate that for the single-frequency estimation problem, the use of nonuniform sampling intervals does not affect the ability to estimate a frequency. However, for apparently randomly sampled time series, aliases effectively do not exist.

We then proceed to solve the one-frequency-with-Lorentzian-decay problem and discuss a number of surprising implications for how decaying signals should be sampled. Next we examine the two stationary frequency problem in some detail, and demonstrate that (1) the ability to estimate two close frequencies is essentially independent of the separation as long as that separation is at least one Nyquist step $|\omega_1 - \omega_2| \geq 2\pi/N$; and (2) that these frequencies are still resolvable at separations corresponding to less than one half step, where the discrete Fourier transform shows only a single peak.

After the two-frequency problem we discuss briefly the multiple nonstationary frequency estimation problem. In Chapter 3 Eq. (3.17) we derive the joint posterior probability of multiple stationary or nonstationary frequencies independent of their amplitude and phase and independent of the noise variance. Here we investigate some of the implications of these formulas and discuss the techniques and procedures needed to apply them effectively.

In Chapter 7, we apply the theory to a number of real time series, including Wolf's relative sunspot numbers, some NMR (nuclear magnetic resonance) data containing multiple close frequencies with decay, and to economic time series which have large trends. The most spectacular results obtained to date are with NMR data, because here prior information tells us very accurately what the "true" model must be.

Equally important, particularly in economics, is the way probability theory deals with trend. Instead of seeking to eliminate the trend from the data (which is known to introduce spurious artifacts that distort the information in the data), we seek instead to eliminate the effect of trend from the final conclusions, leaving the data intact. This proves to be not only a safer, but also a more powerful procedure than detrending the data. Indeed, it is now clear that many published economic time series have been rendered nearly useless because the data have been detrended or seasonally adjusted in an irreversible way that destroys information which probability theory could have extracted from the raw, unmutilated data.

In the last example we investigate the use of multiple measurements and show that probability theory can continue to obtain the standard $\sqrt{n}$ improvement in parameter estimates under much wider conditions than averaging. The analyses presented in Chapter 7 will give the reader a better feel for the types of applications and complex phenomena which can be investigated easily using Bayesian techniques.

1.1 Historical Perspective

Comprehensive histories of the spectral analysis problem have been given recently by Robinson [2] and Marple [3]. We sketch here only the part of it that is directly ancestral to the new work reported here. The problem of determining a frequency in time sampled data is very old; the first astronomers were trying to solve this problem when they attempted to determine the length of a year or the period of the moon. Their methods were crude and consisted of little more than trying to locate the maxima or the nodes of an approximately periodic function. The first significant advance in the frequency estimation problem occurred in the early nineteenth century, when two separate methods of analyzing the problem came into being: the use of probability theory, and the use of the Fourier transform.

Probabilistic methods of dealing with the problem were formulated in some generality by Laplace [4] in the late 18th century, and then applied by Legendre and Gauss [5] [6] who first used (or at least first published) the method of least squares to estimate model parameters in noisy data. In this procedure some idealized model signal is postulated and the criterion of minimizing the sum of the squares of the "residuals" (the discrepancies between the model and the data) is used to estimate the model parameters. In the problem of determining a frequency, the model might be a single cosine with an amplitude, phase, and frequency, contaminated by noise with an unknown variance. Generally one is not interested in the amplitude, phase, or noise variance; ideally one would like to formulate the problem in such a way that only the frequency remains, but this is not possible with direct least squares, which requires us to fit all the model parameters. The method of least squares may be difficult to use in practice; in principle it is well understood. In the case of Gaussian noise, the least squares estimates are simply the parameter values that maximize the probability that we would obtain the data, if a model signal was present with those parameters.

The spectral method of dealing with this problem also has its origin in the early part of the nineteenth century. The Fourier transform is one of the most powerful tools in analysis, and its discrete analogue is by definition the spectrum of the time sampled data. How this is related to the spectrum of the original time series is, however, a nontrivial technical problem whose answer is different in different circumstances. Using the discrete Fourier transform of the data as an estimate of the "true" spectrum is, intuitively, a natural thing to do: after all, the discrete Fourier transform is the spectrum of the noisy time sampled series, and when the noise goes away the discrete Fourier transform is the spectrum of the sampled "true" series, but calculating the spectrum of a series and estimating a frequency are very different problems. One of the things we will attempt to do is to do is to exhibit the exact conditions under which the discrete Fourier transform is an optimal frequency estimator.

With the introduction (or rather, rediscovery [7], [8], [9]) of the fast Fourier transform by Cooley and Tukey [10] in 1965 and the development of computers, the use of the discrete Fourier transform as a frequency and power spectral estimator has become very commonplace. Like the method of least squares, the use of discrete Fourier transform as a frequency estimator is well understood. If the data consist of a signal plus noise, then by linearity the Fourier transform will be the signal transform plus a noise transform. If one has plenty of data the noise transform will be, usually, a function of frequency with slowly varying amplitude and rapidly varying phase. If the peak of the signal transform is larger than the noise transform, the added noise does not change the location of the peak very much. One can then estimate the frequency from the location of the peak of the data transform, as intuition suggests.

Unfortunately, this technique does not work well when the signal-to-noise ratio of the data is small; then we need probability theory. The technique also has problems when the signal is other than a simple harmonic frequency: then the signal has some type of structure [for example Lorentzian or Gaussian decay, or chirp: a chirped signal has the form $\cos(\theta + \omega t + \alpha t^2)$]. The peak will then be spread out relative to a simple harmonic spectrum. This allows the noise to interfere with the parameter estimation problem much more severely, and probability theory becomes essential. Additionally, the Fourier transform is not well defined when the data are nonuniform in time, even though the problem of frequency estimation is not essentially changed.

Arthur Schuster [11] introduced the periodogram near the beginning of this century, merely as an intuitive *ad hoc* method of detecting a periodicity and estimating its frequency. The periodogram is essentially the squared magnitude of the discrete Fourier transform of the data $D \equiv \{d_1, d_2, \cdots, d_N\}$ and can be defined as

$$C(\omega) = \frac{1}{N}\left[R(\omega)^2 + I(\omega)^2\right] = \frac{1}{N}\left|\sum_{j=1}^{N} d_j e^{i\omega t_j}\right|^2, \tag{1.1}$$

where $R(\omega)$, and $I(\omega)$ are the real and imaginary parts of the sum [Eqs. (2.4), and (2.5) below], and N is the total number of data points. The periodogram remains well defined when the frequency ω is allowed to vary continuously or when the data are nonuniform. This avoids one of the potential drawbacks of using this method but does not aid in the frequency estimation problem when the signal is not stationary. Although Schuster himself had very little success with it, more recent experience has shown that regardless of its drawbacks, indeed the discrete Fourier transform or the periodogram does yield useful frequency estimates under a wide variety of conditions. Like least squares, Fourier analysis alone does not give an indication of the accuracy of the estimates of spectral density, although the width of a sharp peak is suggestive of the accuracy of determination of the position of a very sharp line.

In the 160 years since the introduction of the spectral and probability theory methods no particular connection between them had been noted, yet each of these methods seems to function well in some conditions. That these methods could be very closely related (from some viewpoints essentially the same) was shown when Jaynes [12] derived the periodogram directly from the principles of probability theory and demonstrated it to be, a "sufficient statistic" for inferences about a single stationary frequency or "signal" in a time sampled data set, when a Gaussian probability distribution is assigned for the noise. That is, starting with the same probability distribution for the noise that had been used for maximum likelihood or least squares, the periodogram was shown to be the only function of the data needed to make estimates of the frequency; i.e. it summarizes all the information in the data that is relevant to the problem.

In this work we will continue the analysis started by Jaynes and show that when the noise variance σ^2 is known, the conditional posterior probability density of a frequency ω given the data D, the noise variance σ^2, and the prior information I is simply related to the periodogram:

$$P(\omega|D, \sigma, I) \propto \exp\left\{\frac{C(\omega)}{\sigma^2}\right\}. \tag{1.2}$$

Thus, we will have demonstrated the relation between the two techniques. Because the periodogram, and therefore the Fourier transform, will have been derived from the principles of probability theory we will be able to see more clearly under what conditions the discrete Fourier transform of the data is a valid frequency estimator and the proper way to extract optimum estimates from it. Also, from (1.2) we will be able to assess the accuracy of our estimates, which neither least squares, Fourier analysis, nor maximum likelihood give directly.

The term "spectral analysis" has been used in the past to denote a wider class of problems than we shall consider here; often, one has taken the view that the entire time series is a "stochastic process" with an intrinsically continuous spectrum, which we seek to infer. This appears to have been the viewpoint underlying the work of Schuster, and of Blackman-Tukey noted in the following sections. For an account of the large volume of literature on this version of the spectral estimation problem, we refer the reader to Marple [3].

The present work is concerned with what Marple calls the "parameter estimation method". Recent experience has taught us that this is usually a more realistic way of looking at current applications; and that when the parameter estimation approach is based on a correct model it can achieve far better results than can a "stochastic" approach, because it incorporates cogent prior information into the calculation. In addition, the parameter estimation approach proves to be more flexible in ways that are important in applications, adapting itself easily to such complicating features as chirp, decay, or trend.

1.2 Method of Calculation

The basic reasoning used in this work will be a straightforward application of Bayes' theorem: denoting by $P(A|B)$ the conditional probability that proposition A is true, given that proposition B is true, Bayes' theorem is

$$P(H|D, I) = \frac{P(H|I)P(D|H, I)}{P(D|I)}. \tag{1.3}$$

It is nothing but the probabilistic statement of an almost trivial fact: Aristotelian logic is commutative. That is, the propositions

$$HD = \text{" Both } H \text{ and } D \text{ are true"}$$

$$DH = \text{"Both } D \text{ and } H \text{ are true"}$$

say the same thing, so they must have the same truth value in logic and the same probability, whatever our information about them. In the product rule of probability theory, we may then interchange H and D

$$P(H, D|I) = P(D|I)P(H|D, I) = P(H|I)P(D|H, I)$$

which is Bayes' theorem. In our problems, H is any hypothesis to be tested, D is the data, and I is the prior information. In the terminology of the current statistical literature, $P(H|D, I)$ is called the posterior probability of the hypothesis, given the data and the prior information. This is what we would like to compute for several different hypotheses concerning what systematic "signal" is present in our data. Bayes' theorem tells us that to compute it we must have three terms: $P(H|I)$ is the prior probability of the hypothesis (given only our prior information), $P(D|I)$ is the prior probability of the data (this term will always be absorbed into a normalization constant and will not change the conclusions within the context of a given model, although it does affect the relative probabilities of different models) and $P(D|H, I)$ is called the direct probability of the data, given the hypothesis and the prior information. The direct probability is called the "sampling distribution" when the hypothesis is held constant and one considers different sets of data, and it is called the "likelihood function" when the data are held constant and one varies the hypothesis. Often, a prior probability distribution is called simply a "prior".

In a specific Bayesian probability calculation, we need to "define our model"; i.e. to enumerate the set $\{H_1, H_2, \cdots\}$ of hypotheses concerning the systematic signal in the model, that is to be tested by the calculation. A serious weakness of all Fourier transform methods is that they do not consider this aspect of the problem. In the widely used Blackman-Tukey [13] method of spectrum analysis, for example, there is no mention of any model or any systematic signal at all. In the problems we are considering, specification of a definite model (i.e. stating just what prior information we have about the phenomenon being observed) is essential; the information we can extract from the data depends crucially on which model we analyze.

In our problems, therefore, the Blackman-Tukey method, which does not even have the concept of a signal, much less a signal-to-noise ratio, would be inappropriate. Bayesian analysis based on a good model can achieve orders of magnitude better sensitivity and resolution. Indeed, one of our main new results is the very great improvement in resolution that can be achieved by replacing an unrealistic model by a realistic one.

In the most general model we will analyze, the hypothesis H will be of the form

$$H \equiv \text{“}f(t) = \sum_{j=1}^{m} B_j G_j(t, \{\omega\})\text{”}$$

where $f(t)$ is some analytic representation of the time series, $G_j(t, \{\omega\})$ is one particular model function (for example a sinusoid or trend), B_j is the amplitude with which G_j enters the model, and $\{\omega\}$ is a set of frequencies, decay rates, chirp rates, trend rate, or any other parameters of interest.

In the problem we are considering we focus our attention on the $\{\omega\}$ parameters. Although we will calculate the expectation value of the amplitudes $\{B\}$ we will not generally be interested in them. We will seek to formulate the posterior probability density $P(\{\omega\}|D, I)$ independently of the amplitudes $\{B\}$.

The principles of probability theory uniquely determine how this is to be done. Suppose ω is a parameter of interest, and B is a "nuisance parameter" that we do not, at least at the moment, need to know. What we want is $P(\omega|D, I)$, the posterior probability (density) of ω. This may be calculated as follows: first calculate the joint posterior probability density of ω and B by Bayes' theorem:

$$P(\omega, B|D, I) = P(\omega, B|I)\frac{P(D|\omega, B, I)}{P(D|I)}$$

and then integrate out B, obtaining the marginal posterior probability density for ω:

$$P(\omega|D, I) = \int dB P(\omega, B|D, I)$$

which expresses what the data and prior information have to tell us about ω, regardless of the value of B.

Although integration over the nuisance parameters may look a little strange at first glance, it is easily demonstrated to be a straightforward application of the sum rule of probability theory: the probability that one of several mutually exclusive propositions is true, is the sum of their separate probabilities. Suppose for simplicity that B is a discrete variable taking on the values $\{B_1, \cdots, B_n\}$ and we know that when the data were taken only one value of B was realized; but we do not know which value. We can compute $P(\omega, \sum_{i=1}^{n} B_i|D, I)$ where the symbol "+" or "$\sum$" inside a probability symbol means the Boolean "or" operation [read this as the probability of (ω and B_1) or (ω and B_2)$\cdots$]. Using the sum rule this probability may be written

$$\begin{aligned} P(\omega, B_1 + \sum_{i=2}^{n} B_i|D, I) &= P(\omega, B_1|D, I) \\ &+ P(\omega, \sum_{i=2}^{n} B_i|D, I)[1 - P(\omega, B_1|\sum_{i=2}^{n} B_i D, I)]. \end{aligned}$$

The last term $P(\omega, B_1|\sum_{i=2}^{n} B_i D, I)$ is zero: only one value of B could be realized. We have

$$P(\omega, B_1 + \sum_{i=2}^{n} B_i|D, I) = P(\omega, B_1|D, I) + P(\omega, \sum_{i=2}^{n} B_i|D, I)$$

and repeated application of the sum rule gives

$$P(\omega, \sum_{i=1}^{n} B_i|D, I) = \sum_{i=1}^{n} P(\omega, B_i|D, I).$$

When the values of B are continuous the sums go into integrals and one has

$$P(\omega|D, I) = \int dB P(\omega, B|D, I), \tag{1.4}$$

the given rule. The term on the left is called the marginal posterior probability density function of ω, and it takes into account all possible values of B regardless of which actual value was realized. We have dropped the reference to B specifically because this distribution no longer depends on one specific value of B; it depends rather on all of them.

We discuss these points further in Appendices A, B, and C where we show that this procedure is similar to, but superior to, the common practice of estimating the parameter B from the data and then constraining B to that estimate.

In the following chapter we consider the simplest nontrivial spectral estimation model

$$f(t) = B_1 \cos \omega t + B_2 \sin \omega t$$

and analyze it in some depth to show some elementary but important points of principle in the technique of using probability theory with nuisance parameters and "uninformative" priors.

Chapter 2

SINGLE STATIONARY SINUSOID PLUS NOISE

2.1 The Model

We begin the analysis by constructing the direct probability, $P(D|H,I)$. We think of this as the likelihood of the parameters, because it is its dependence on the model parameters which concerns us here. The time series $y(t)$ we are considering is postulated to contain a single stationary harmonic signal $f(t)$ plus noise $e(t)$. The basic model is always: we have recorded a discrete data set $D = \{d_1, \cdots, d_N\}$; sampled from $y(t)$ at discrete times $\{t_1, \cdots, t_N\}$; with a model equation

$$d_i = y(t_i) = f(t_i) + e_i, \qquad (1 \leq i \leq N).$$

As already noted, *different models correspond to different choices of the signal $f(t)$.* We repeat the analysis originally done by Jaynes [12] using a different, but equivalent, set of model functions. We repeat this analysis for three reasons: first, by using a different formulation of the problem we can see how to generalize to multiple frequencies and more complex models; second, to introduce a different prior probability for the amplitudes, which simplifies the calculation but has almost no effect on the final result; and third, to introduce and discuss the calculation techniques without the complex model functions confusing the issues.

The model for a simple harmonic frequency may be written

$$f(t) = B_1 \cos(\omega t) + B_2 \sin(\omega t) \tag{2.1}$$

which has three parameters (B_1, B_2, ω) that may be estimated from the data. The model used by Jaynes [12] was the same, but expressed in polar coordinates:

$$f(t) = B\cos(\omega t + \theta)$$

$$B = \sqrt{B_1^2 + B_2^2}$$

$$\tan\theta = -\frac{B_2}{B_1}$$

$$dB_1 dB_2 d\omega = BdBd\theta d\omega.$$

It is the factor B in the volume elements which is treated differently in the two calculations. Jaynes used a prior probability that initially considered equal intervals of θ and B to be equally likely, while we shall use a prior that initially considers equal intervals of B_1 and B_2 to be equally likely.

Of course, neither choice fully expresses all the prior knowledge we are likely to have in a real problem. This means that the results we find are conservative, and in a case where we have quite specific prior information about the parameters, we would be able to do somewhat better than in the following calculation. However, the differences arising from different prior probabilities are small provided we have a reasonable amount of data. For a detailed discussion and derivation of the prior probabilities used in this chapter, see Appendix A. In addition, in Appendix D we show explicitly that the prior used by Jaynes is more conservative for frequency estimation than the uniform prior we use, but when the signal-to-noise ratio is large the effect of this uninformative prior is completely negligible.

2.2 The Likelihood Function

To construct the likelihood we take the difference between the model function, or "signal", and the data. If we knew the true signal, then this difference would be just the noise. Then if we knew the probability of the noise we could compute the direct probability or likelihood. We wish to assign a noise prior probability density which is consistent with the available information about the noise. The prior should be as uninformative as possible to prevent us from "seeing" things in the data which are not there.

To derive the prior probability for the noise is a problem that can be approached in various ways. Perhaps the most general one is to view it as a simple application of the principle of maximum entropy. Let $P(e|I)$ stand for the probability that the noise has value "e" given the prior information I. Then, assuming the second moment of the noise (i.e. the noise power) is known, the entropy functional which must be maximized is given by

$$-\int_{-\infty}^{+\infty} P(e|I)\log P(e|I)de - \lambda\int_{-\infty}^{+\infty} e^2 P(e|I)de - \beta\int_{-\infty}^{+\infty} P(e|I)de$$

where λ is the Lagrange multiplier associated with the second moment, and β is the multiplier for normalization. The solution to this standard maximization problem is

$$P(e|\lambda, I) = (\lambda/\pi)^{\frac{1}{2}}\exp\left\{-\lambda e^2\right\}.$$

Adopting the notation $\lambda = (2\sigma^2)^{-1}$, where σ^2 is the second moment, assumed known, we have

$$P(e|\sigma, I) = \frac{1}{\sqrt{2\pi\sigma^2}}\exp\left\{-\frac{e^2}{2\sigma^2}\right\}.$$

This is a Gaussian distribution, and when σ is taken as the RMS noise level, it is the least informative prior probability density for the noise that is consistent with the given second moment. By least informative we mean that: if any of our assumptions had been different and we used that information in maximum entropy to derive a new prior probability for the noise, then for a given σ, that new probability density would be less spread out, thus our accuracy estimates would be narrowed. Thus, in the calculations below, we will be claiming less accuracy than would be justified had we included those additional effects in deriving the prior probability for the noise. The point is discussed further in Chapter 5. In Chapter 6 we demonstrate (numerically) the effects of violating the assumptions that will go into the calculation. All of these "conservative" features are safety devices which make it impossible for the theory to mislead us by giving overoptimistic results.

Having the prior probability for the noise, and adopting the notation: e_i is the noise at time t_i, we apply the product rule of probability theory to obtain the probability that we would obtain a set of noise values $\{e_1, \cdots, e_N\}$: supposing the e_i independent in the sense that $P(e_i|e_j, \sigma, I) = P(e_i|\sigma, I)$ this is given by

$$P(e_1, \cdots, e_N|\sigma, I) \propto \prod_{i=1}^{N}\left[\frac{1}{\sqrt{2\pi\sigma^2}}\exp(-\frac{e_i^2}{2\sigma^2})\right],$$

in which the independence of different e_i is also a safety device to maintain the conservative aspect. But if we have definite prior evidence of dependence, i.e. correlations, it is a simple computational detail to take it into account as noted later.

Other rationales for this choice exist in other situations. For example, if the noise is known to be the result of many small independent effects, the central limit theorem of probability theory leads to the Gaussian form independently of the fine details, even if the second moment is not known. For a detailed discussion of why and when a Gaussian distribution should be used for the noise probability, see the original paper by Jaynes [12]. Additionally, the book of Jaynes' collected papers contains a discussion of the principle of maximum entropy and much more [14].

If we have the true model, the difference between the data d_i and the model f_i is just the noise. Then the direct probability that we should obtain the data $D = \{d_1 \cdots d_N\}$, given the parameters, is proportional to the likelihood function:

$$P(D|B_1, B_2, \omega, \sigma, I) \propto L(B_1, B_2, \omega, \sigma) = \prod_{i=1}^{N} \sigma^{-1} \exp\{-\frac{1}{2\sigma^2}[d_i - f(t_i)]^2\}$$

$$L(B_1, B_2, \omega, \sigma) = \sigma^{-N} \times \exp\{-\frac{1}{2\sigma^2}\sum_{i=1}^{N}[d_i - f(t_i)]^2\}. \tag{2.2}$$

The usual way to proceed is to fit the sum in the exponent. Finding the parameter values which minimize this sum is called "least squares". The equivalent procedure (in this case) of finding parameter values that maximize $L(B_1, B_2, \omega, \sigma)$ is called "maximum likelihood". The maximum likelihood procedure is more general than least squares: it has theoretical justification when the likelihood is not Gaussian. The departure of Jaynes was to use (2.2) in Bayes' theorem (1.3), and then to remove the phase and amplitude from further consideration by integration over these parameters.

In doing this preliminary calculation we will make a number of simplifying assumptions, then in Chapter 3 correct them by solving a much more general problem exactly. For now we insert the model (2.1) into the likelihood (2.2) and expand the exponent to obtain:

$$L(B_1, B_2, \omega, \sigma) \propto \sigma^{-N} \exp\left\{-\frac{NQ}{2\sigma^2}\right\} \tag{2.3}$$

where

$$Q \equiv \overline{d^2} - \frac{2}{N}[B_1 R(\omega) + B_2 I(\omega)] + \frac{1}{2}(B_1^2 + B_2^2),$$

and

$$R(\omega) = \sum_{i=1}^{N} d_i \cos(\omega t_i) \tag{2.4}$$

$$I(\omega) = \sum_{i=1}^{N} d_i \sin(\omega t_i) \tag{2.5}$$

are the functions introduced in (1.1), and

$$\overline{d^2} = \frac{1}{N}\sum_{i=1}^{N} d_i^2$$

is the observed mean-square data value. In this preliminary discussion we assumed the data have zero mean value (any nonzero average value has been subtracted from the data), and we simplified the quadratic term as follows:

$$\sum_{i=1}^{N} f(t_i)^2 = B_1^2 \sum_{i=1}^{N} \cos^2 \omega t_i + B_2^2 \sum_{i=1}^{N} \sin^2 \omega t_i + 2B_1B_2 \sum_{i=1}^{N} \cos(\omega t_i)\sin(\omega t_i),$$

with

$$\sum_{i=1}^{N} \cos^2 \omega t_i = \frac{N}{2} + \frac{1}{2}\sum_{i=1}^{N} \cos 2\omega t_i \simeq \frac{N}{2},$$

$$\sum_{i=1}^{N} \sin^2 \omega t_i = \frac{N}{2} - \frac{1}{2}\sum_{i=1}^{N} \cos 2\omega t_i \simeq \frac{N}{2},$$

$$\sum_{i=1}^{N} \cos(\omega t_i)\sin(\omega t_i) = \frac{1}{2}\sum_{i=1}^{N} \sin(2\omega t_i) \ll \frac{N}{2}$$

so the quadratic term is approximately

$$\sum_{i=1}^{N} f(t_i)^2 \approx \frac{N}{2}\left(B_1^2 + B_2^2\right).$$

The neglected terms are of order one, and small provided $N \gg 1$ (except in the special case $\omega t_N \ll 1$). We will assume, for now, that the data contain no evidence of a low frequency.

The cross term, $\sum_{i=1}^{N} \cos(\omega t_i)\sin(\omega t_i)$, is at most of the same order as the terms we just ignored; therefore, this term is also ignored. The assumption that this cross term is zero is equivalent to assuming the sine and cosine functions are orthogonal on the discrete time sampled set. Indeed, this is the actual case for uniformly spaced time intervals; however, even without uniform spacing this is a good approximation provided N is large. The assumption that the cross terms are zero by orthogonality will prove to be the key to generalizing this problem to more complex models, and in Chapter 3 the assumptions that we are making now will become exact by a change of variables.

2.3 Elimination of Nuisance Parameters

In a harmonic analysis one is primarily interested in the frequency ω. Then if the amplitude, phase, and the variance of the noise are unknown, they are nuisance parameters. We gave the general procedure for dealing with nuisance parameters in Chapter 1. To apply that rule we must integrate the posterior probability density with respect to B_1, B_2, and also σ if the noise variance is unknown.

If we had prior information about the nuisance parameters (such as: they had to be positive, they could not exceed an upper limit, or we had independently measured values for them) then here would be the place to incorporate that information into the calculation. We illustrate the effects of integrating over a nuisance parameter, as well as the use of prior information in, Appendices B and C and explicitly calculate the expectation values of B_1 and B_2 when a prior measurement is available. At present we assume no prior information about the amplitudes B_1 and B_2 and assign them a prior probability which indicates "complete ignorance of a location parameter". This prior is a uniform, flat, prior density; it is called an improper prior probability because it is not normalizable. In principle, we should approach an improper prior as the limit of a sequence of proper priors. The point is discussed further in Appendices A and B. However, in this problem there are no difficulties with the use of the uniform prior because the Gaussian cutoff in the likelihood function ensures convergence in (2.3).

Upon multiplying the likelihood (2.3) by the uniform prior and integrating with respect to B_1 and B_2 one obtains the joint quasi-likelihood of ω and σ:

$$L(\omega, \sigma) \propto \sigma^{-N+2} \times \exp\left\{-\frac{N}{2\sigma^2}[\overline{d^2} - 2C(\omega)/N]\right\} \tag{2.6}$$

where $C(\omega)$, the Schuster periodogram defined in (1.1), has appeared in a very natural and unavoidable way. If one knows the variance σ^2 from some independent source and has no additional prior information about ω, then the problem is completed. The posterior probability density for ω is given by

$$P(\omega|D, \sigma, I) \propto \exp\left\{\frac{C(\omega)}{\sigma^2}\right\}. \tag{2.7}$$

Because we have assumed little prior information about B_1, B_2, ω and have made conservative assumptions about the noise; this probability density will yield conservative estimates of ω. By this we mean, as before, that if we had more prior information, we could exploit it to obtain still better results. We will illustrate this point further in Chapter 5 and show that when the data have characteristics which differ from our assumptions, Eq. (2.7) will always make a conservative estimates of the frequency ω. Thus the assumptions we are making act as safeguards to protect us from seeing things in the data that are not really there. We place such great stress on this point because we shall presently obtain some surprisingly sharp estimates.

The above analysis is valid whenever the noise variance (or power) is known. Frequently one has no independent prior knowledge of the noise. The noise variance σ^2 then becomes a nuisance parameter. We eliminate it in much the same way as the amplitudes were eliminated. Now σ is restricted to positive values and additionally it is a scale parameter. The prior which indicates "complete ignorance" of a scale is the Jeffreys prior $1/\sigma$ [15]. Multiplying Eq. (2.6) by the Jeffreys prior and integrating over all positive values gives

$$P(\omega|D,I) \propto \left[1 - \frac{2C(\omega)}{N\overline{d^2}}\right]^{\frac{2-N}{2}}. \tag{2.8}$$

This is called a "Student t-distribution" for historical reasons, although it is expressed here in very nonstandard notation. In our case it is the posterior probability density that a stationary harmonic frequency ω is present in the data when we have no prior information about σ.

These simple results, Eqs. (2.7) and (2.8), show why the discrete Fourier transform tends to peak at the location of a frequency when the data are noisy. Namely, the discrete Fourier transform is directly related to the probability that a single harmonic frequency is present in the data, even when the noise level is unknown. Additionally, zero padding a time series (i.e. adding zeros at its end to make a longer series) and then taking the Fast Fourier transform of the padded series, is equivalent to calculating the Schuster periodogram at smaller frequency intervals. If the signal one is analyzing is a simple harmonic frequency plus noise, then the maximum of the periodogram will be the "best" estimate of the frequency that we can make in the absence of additional prior information about it.

We now see the discrete Fourier transform and the Schuster periodogram in a entirely new light: the highest peak in the discrete Fourier transform is an optimal frequency estimator for a data set which contains a single harmonic frequency in the presence of Gaussian white noise. Stated more carefully, the discrete Fourier transform will give optimal frequency estimates if six conditions are met:

1. The number of data values N is large,
2. There is no constant component in the data,
3. There is no evidence of a low frequency,
4. The data contain only one frequency,
5. The frequency must be stationary
   (i.e. the amplitude and phase are constant),
6. The noise is white.

If any of these six conditions is not met, the discrete Fourier transform may give misleading or simply incorrect results in light of the more realistic models. Not because the discrete Fourier transform is wrong, but because it is answering what we should regard as the wrong question. The discrete Fourier transform will always interpret the data in terms of a single harmonic frequency model! In Chapter 6 we illustrate the effects of violating one or more of these assumptions and demonstrate that when they are violated the estimated parameters are always less certain than when these conditions are met.

2.4 Resolving Power

When the six conditions are met, just how accurately can the frequency be estimated? This question is easily answered; we do this by approximating (2.7) by a Gaussian and then making the (mean) $\pm$ (standard deviation) estimates of the frequency ω. Expanding $C(\omega)$ about the maximum $\hat{\omega}$ we have

$$C(\omega) = C(\hat{\omega}) - \frac{b}{2}(\hat{\omega} - \omega)^2 + \cdots$$

where

$$b \equiv -C''(\hat{\omega}) > 0. \tag{2.9}$$

The Gaussian approximation is

$$P(\omega|D,\sigma,I) \simeq \left[\frac{2b}{\pi\sigma^2}\right]^{\frac{1}{2}} \exp\left\{-\frac{b(\hat{\omega}-\omega)^2}{2\sigma^2}\right\}$$

from which we would make the (mean) $\pm$ (standard deviation) estimate of the frequency

$$\omega_{\text{est}} = \hat{\omega} \pm \frac{\sigma}{\sqrt{b}}.$$

The accuracy depends on the curvature of $C(\omega)$ at its peak, not on the height of $C(\omega)$. For example, if the data are composed of a single sine wave plus noise $e(t)$ of standard deviation σ

$$d_t = \hat{B}_1 \cos(\hat{\omega} t) + e_t$$

then as found by Jaynes [12]:

$$C(\omega_{\max}) \simeq \frac{N\hat{B}_1^2}{4}$$

$$b \simeq \frac{\hat{B}_1^2 N^3}{48}$$

$$(\omega)_{\text{est}} = \hat{\omega} \pm \frac{\sigma}{|B_1|}\sqrt{48/N^3} \tag{2.10}$$

which indicates, as intuition would lead us to expect, that the accuracy depends on the signal-to-noise ratio, and quite strongly on how much data we have.

The height of the posterior probability density increases like the exponential of $N\hat{B}_1^2/4\sigma^2$ while the error estimates depend on the exponential of $N^3\hat{B}_1^2/96\sigma^2$. If one has a choice between doubling the amount of data $N \rightarrow 2N$, or doubling the signal-to-noise ratio $\hat{B}_1/\sigma \rightarrow 2\hat{B}_1/\sigma$, always double the amount of data if you have detected the signal, and always double the signal-to-noise ratio if you have no strong evidence of a signal.

If we have sufficient signal-to-noise ratio for the posterior probability density $\exp\{N\hat{B}_1^2/4\sigma^2\}$ to have a peak well above the noise, doubling the amount of data, $N \rightarrow 2N$ will double the height of the periodogram giving $\exp\{N\hat{B}_1^2/4\sigma^2\}$ times more evidence of a frequency while the error will go down like $\sqrt{8}$. On the other hand, if the signal-to-noise ratio is so low that $\exp\{N\hat{B}_1^2/4\sigma^2\}$ has no clear peak above the noise, then doubling the signal-to-noise ratio $\hat{B}_1^2/\sigma^2 \rightarrow 4\hat{B}_1^2/\sigma^2$ will give $\exp\{3N\hat{B}_1^2/4\sigma^2\}$ times more evidence for a frequency, while the error goes down by 2. The trade off is clear: if you have sufficient signal-to-noise for signal detection more data are important for resolution; otherwise more signal-to-noise will detect the signal with less data.

We can further compare these results with experience, but first note that we are using dimensionless units, since we took the data sampling interval to be 1. Converting to ordinary physical units, let the sampling interval be Δt seconds, and denote by f the frequency in Hz. Then the total number of cycles in our data record is

$$\frac{\hat{\omega}(N-1)}{2\pi} = (N-1)\hat{f}\Delta t = \hat{f}T$$

where $T = (N-1)\Delta t$ seconds is the duration of our data run. So the conversion of dimensionless ω to f in physical units is

$$f = \frac{\omega}{2\pi\Delta t} \text{ Hz.}$$

The frequency estimate (2.10) becomes

$$f_{\text{est}} = \hat{f} \pm \delta f \text{ Hz}$$

where now, not distinguishing between N and $(N-1)$,

$$\delta f = \frac{\sigma}{2\pi\hat{B}_1 T}\sqrt{48/N} = \frac{1.1\sigma}{\hat{B}_1 T\sqrt{N}} \text{ Hz.} \tag{2.11}$$

Comparing this with (2.10) we now see that to improve the accuracy of the estimate the two most important factors are how long we sample (the T dependence) and the signal-to-noise ratio. We could double the number of data values in one of two ways, by doubling the total sampling time or by doubling the sampling rate. However, (2.11) clearly indicates that doubling the sampling time is to be preferred. This indicates that data values near the beginning and end of a record are most important for frequency estimation, in agreement with intuitive common sense.

Let us take a specific example: if we have an RMS signal-to-noise ratio (i.e. ratio of RMS signal to RMS noise $\equiv$ S/N) of S/N $= \hat{B}_1/\sqrt{2}\sigma = 1$, and we take data every $\Delta t = 10^{-3}$ sec. for $T = 1$ second, thus getting $N = 1000$ data points, the theoretical accuracy for determining the frequency of a single steady sinusoid is

$$\delta f = \frac{1.1}{\sqrt{2000}} = 0.025 \text{ Hz} \tag{2.12}$$

while the Nyquist frequency for the onset of aliasing is $f_N = (2\Delta t)^{-1} = 500$Hz, greater by a factor of 20,000.

To some, this result will be quite startling. Indeed, had we considered the periodogram itself to be a spectrum estimator, we would have calculated instead the width of its central peak. A noiseless sinusoid of frequency $\hat{\omega}$ would have a periodogram proportional to

$$C(\omega) \propto \frac{\sin^2\{N(\omega-\hat{\omega})/2\}}{\sin^2\{(\omega-\hat{\omega})/2\}}$$

thus the half-width at half amplitude is given by $|N(\hat{\omega}-\omega)/2| = \pi/4$ or $\delta\omega = \pi/2N$. Converting to physical units, the periodogram will have a width of about

$$\delta f = \frac{1}{4N\Delta t} = \frac{1}{4T} = 0.25 \text{ Hz} \tag{2.13}$$

just ten times greater than the value (2.12) indicated by probability theory. This factor of ten is the amount of narrowing produced by the exponential peaking of the periodogram in (2.7), even for unit signal-to-noise ratio.

But some would consider even the result (2.13) to be a little overoptimistic. The famous Rayleigh criterion [16] for resolving power of an optical instrument supposes that the minimum resolvable frequency difference corresponds to the peak of the periodogram of one sinusoid coming at the first zero of the periodogram of the second. This is twice (2.13):

$$\delta f_{\text{Rayleigh}} = \frac{1}{2T} = 0.5 \text{ Hz}. \tag{2.14}$$

There is a widely believed "folk-theorem" among theoreticians without laboratory experience, which seems to confuse the Rayleigh limit with the Heisenberg uncertainty principle, and holds that (2.14) is a fundamental irreducible limit of resolution. Of course there is no such theorem, and workers in high resolution NMR have been routinely determining line positions to an accuracy that surpasses the Rayleigh limit by an order of magnitude, for thirty years.

The misconception is perhaps strengthened by the curious coincidence that (2.14) is also the minimum half-width that can be achieved by a Blackman-Tukey spectrum analysis [13] (even at infinite signal-to-noise ratio) because the "Hanning window" tapering function that is applied to the data to suppress side-lobes (the secondary maxima of $[\sin(x)/x]^2$) just doubles the width of the periodogram. Since the Blackman-Tukey method has been used widely by economists, oceanographers, geophysicists, and engineers for many years, it has taken on the appearance of an optimum procedure.

According to E.T. Jaynes, Tukey himself acknowledged [17] that his method fails to give optimum resolution, but held this to be of no importance because "real time series do not have sharp lines." Nevertheless, this misconception is so strongly held that there have been attacks on the claims of Bayesian/Maximum Entropy spectrum analysts to be able to achieve results like (2.12) when the assumed conditions are met. Some have tried to put such results in the same category with circle squaring and perpetual motion machines. Therefore we want to digress to explain in very elementary physical terms why it is the Bayesian result (2.11) that does correspond to what a skilled experimentalist can achieve.

Suppose first that our only data analysis tool is our own eyes looking at a plot of the raw data of duration $T = 1$ sec., and that the unknown frequency f in (2.12) is 100Hz. Now anyone who has looked at a record of a sinusoid and equal amplitude wide-band noise, knows that the cycles are quite visible to the eye. One can count the total number of cycles in the record confidently (using interpolation to help us over the doubtful regions) and will feel quite sure that the count is not in error by even one cycle. Therefore by raw eyeballing of the data and counting the cycles, one can achieve an accuracy of

$$\delta f \simeq \frac{1}{T} = 1 \text{ Hz}.$$

But in fact, if one draws the sine wave that seems to fit the data best, he can make a quite reliable estimate of how many quarter-cycles were in the data, and thus achieve

$$\delta f \simeq \frac{1}{4T} = 0.25 \text{ Hz}$$

corresponding just to the periodogram width (2.13).

Then the use of probability theory needs to surpass the naked eye by another factor of ten to achieve the Bayesian width (2.12). What probability theory does is essentially to average out the noise in a way that the naked eye cannot do. If we repeat some measurement N times, any randomly varying component of the data will be suppressed relative to the systematic component by a factor of $N^{-\frac{1}{2}}$, the standard rule.

In the case considered, we assumed $N = 1000$ data points. If they were all independent measurements of the same quantity with the same accuracy, this would suppress the noise by about a factor of 30. But in our case not all measurements are equally cogent for estimating the frequency. Data points in the middle of the record contribute very little to the result; only data points near the ends are highly relevant for determining the frequency, so the effective number of observations is less than 1000. The probability analysis leading to (2.12) indicates that the "effective number of observations" is only about $N/10 = 100$; thus the Bayesian width (2.12) that results from the exponential peaking of the periodogram now appears to be, if anything, somewhat conservative.

Indeed, that is what Bayesian analysis always does when we use smooth, uninformative priors for the parameters, because then probability theory makes allowance for all possible values that they might have. As noted before, if we had any cogent prior information about ω and expressed it in a narrower prior, we would be led to still better results; but they would not be much better unless the prior range became comparable to the width of the likelihood $L(\omega)$.

2.5 The Power Spectral Density $\hat{p}$

The usual way the result from a spectral analysis is displayed is in the form of a power spectral density (i.e. power per unit frequency). In Fourier transform spectroscopy this is typically taken as the squared magnitude of the discrete Fourier transform of the data. We would like to express the results of the present calculation in a similar manner to facilitate comparisons between these techniques, although strictly speaking there is no exact correspondence between a spectral density defined with reference to a stochastic model and one that pertains to a parameter estimation model.

We begin by defining what we mean by the "estimated spectrum," since several quite different meanings of the term can be found in the literature. Define $\hat{p}(\omega)d\omega$ as the expectation, over the joint posterior probability distribution for all the parameters, of the energy carried by the signal (not the noise) in frequency range $d\omega$, during our observation time $t_N - t_1$. Then $\int \hat{p}(\omega)d\omega$ over some frequency range is the expectation of the total energy carried by the signal in that frequency range. The total energy E carried by the signal in our model is

$$E = \int_{t_1}^{t_N} f(t)^2 dt \approx \frac{T}{2}\left(B_1^2 + B_2^2\right)$$

and its expectation is given by

$$\hat{p}(\omega) = \frac{T}{2}\langle B_1^2 + B_2^2\rangle;$$

but $N = T/\Delta t$, where Δt is the sampling time which in dimensionless units is one. The power spectral density is

$$\hat{p}(\omega) = \frac{N}{2}\int dB_1 dB_2 (B_1^2 + B_2^2) P(\omega, B_1, B_2|D, \sigma, I).$$

Performing the integrals over B_1 and B_2 we obtain

$$\hat{p}(\omega) = 2\left[\sigma^2 + C(\omega)\right] P(\omega|D, \sigma, I). \tag{2.15}$$

We see now that the peak of the periodogram is indicative of the total energy carried by the signal. The additional term $2\sigma^2$ is not difficult to explain; but we delay that explanation until after we have derived these results for the general theory (see page 52).

If the noise variance is assumed known, (2.15) becomes

$$\hat{p}(\omega) = 2\left[\sigma^2 + C(\omega)\right]\frac{\exp\{C(\omega)/\sigma^2\}}{\int d\omega \exp\left\{C(\omega)/\sigma^2\right\}}. \tag{2.16}$$

Probability theory will handle those secondary maxima (side lobes) that occur in the periodogram by assigning them negligible weight.

This is easily seen by considering the same example discussed earlier. Take $d(t) = \hat{B}_1\cos(\hat{\omega}t)$ sampled on a uniform grid; then when $\hat{\omega} \simeq \omega$

$$C(\omega) \simeq \frac{\hat{B}_1^2}{4N}\left[\frac{\sin N(\hat{\omega}-\omega)/2}{(\hat{\omega}-\omega)/2}\right]^2$$

and C'' is

$$C'' \equiv b \simeq \frac{\hat{B}_1^2 N^3}{24}$$

and $\hat{p}(\omega)$ is approximately

$$\hat{p}(\omega) \simeq 2\left[\sigma^2 + \frac{4\hat{B}_1^2\sin^2 N(\hat{\omega}-\omega)/2}{(\hat{\omega}-\omega)^2}\right]\left[\frac{\hat{B}_1^2 N^3}{24\pi\sigma^2}\right]^{\frac{1}{2}}\exp\left\{-\frac{\hat{B}_1^2 N^3}{48\sigma^2}(\hat{\omega}-\omega)^2\right\}.$$

Unless the signal-to-noise ratio $\hat{B}_1/\sigma\sqrt{2}$ is very small, this is very nearly a delta function.

If we take $\hat{B}_1 = \sqrt{2}\sigma = 1$, and $N = 1000$ data values, then

$$\hat{p}(\omega) \simeq 2\left[1 + 4\frac{\sin^2 1000(\hat{\omega}-\omega)/2}{(\hat{\omega}-\omega)^2}\right][5150]\exp\{-4\times 10^7(\hat{\omega}-\omega)^2\}.$$

This reaches a maximum value of 10^{11} at $\hat{\omega} = \omega$ and has dropped to $\frac{1}{2}$ this value when $\hat{\omega} - \omega$ has changed by only 0.0001; this function is indeed a good approximation to a delta function and (2.16) may be approximated by:

$$\hat{p}(\omega) \simeq \left[\sigma^2 + C(\hat{\omega})\right][\delta(\hat{\omega}-\omega) + \delta(\hat{\omega}+\omega)]$$

for most purposes. But for the term σ^2, the peak of the periodogram is, in our model, nearly the total energy carried by the signal. It is not an indication of the spectral density as Schuster [11] supposed it to be for a stochastic model. In the present model, the periodogram of the data is not even approximately the spectral energy density of the signal.

2.6 Wolf's Relative Sunspot Numbers

Wolf's relative sunspot numbers are, perhaps, the most analyzed set of data in all of spectrum analysis. As Marple [3] explains in more detail, these numbers (defined as: $W = k[10g + f]$, where g is the number of sunspot groups, f is the number of individual sunspots, and k is used to reduce different telescopes to a common scale) have been collected on a yearly basis since 1700, and on a monthly basis since 1748 [18]. The exact physical mechanism which generates the sunspots is unknown, and no complete theory exists. Different analyses of these numbers have been published more or less regularly since their tabulation began. Here we will analyze the sunspot numbers with a number of different models including the simple harmonic analysis just completed, even though we know this analysis is too simple to be realistic for these numbers.

We have plotted the time series from 1700 to 1985 in Fig. 2.1(A). A cursory examination of this time series does indeed show a cyclic variation with a period of about 11 years. The square of the discrete Fourier transform is a continuous function of frequency and is proportional to the Schuster periodogram of the data Fig. 2.1(B), continuous curve. The frequencies could be restricted to the Nyquist [19] [20] steps (open circles); it is a theorem that the discrete Fourier transform on those points contains all the information that is in the periodogram, but one sees that the information is much more apparent to the eye in the continuous periodogram. The Schuster periodogram or the discrete Fourier transform clearly show a maximum with period near 11 years.

We then compute the "Student t-distribution" (2.8) and have displayed it in figure 2.1(C). Now because of the processing in (2.8) all details in the periodogram have been suppressed and only the peak at 11 years remains.

We determine the accuracy of the frequency estimate as follows: we locate the maximum of the "Student t-distribution", integrate about a symmetric interval, and record the enclosed probability at a number of points. This gives a period of 11.04 years with

period in years		accuracy in years	probability enclosed
11.04	$\pm$	0.015	0.62
	$\pm$	0.020	0.75
	$\pm$	0.026	0.90

Figure 2.1: Wolf's Relative Sunspot Numbers

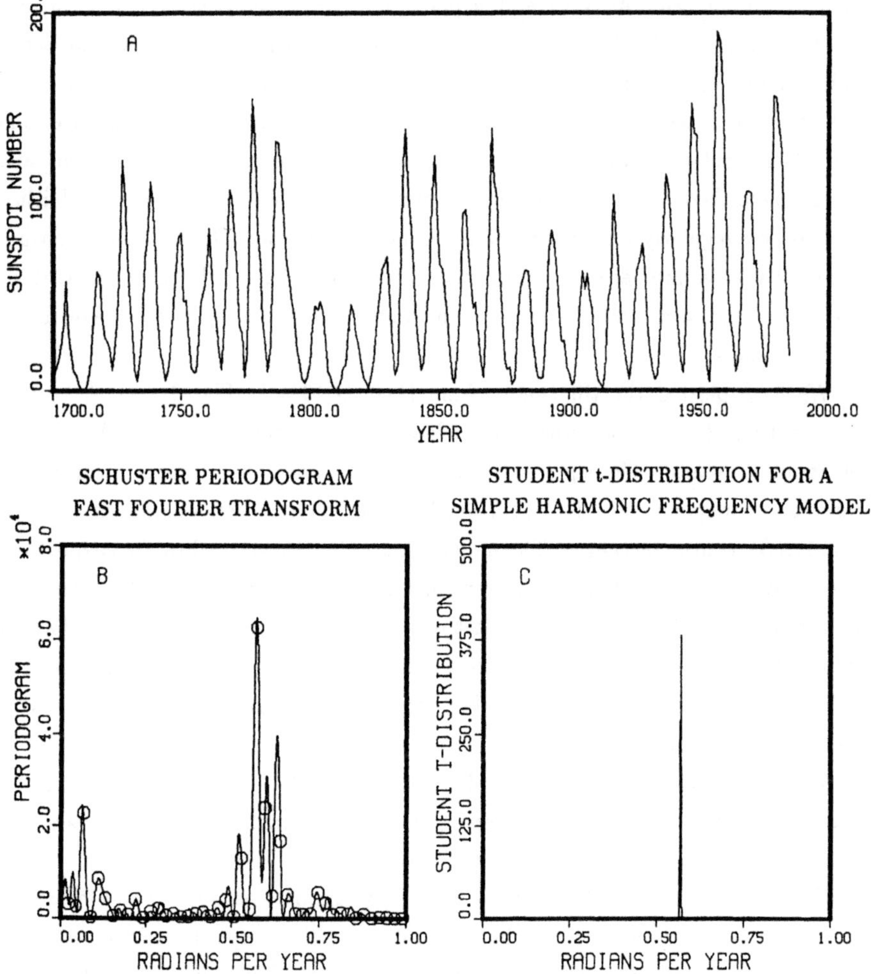

Wolf's relative sunspot numbers (A) have been collected on a yearly basis since 1700. The periodogram (B) contains evidence of several complex phenomena. In spite of this the single frequency model posterior probability density (C) picks out the 11.04 year cycle to an estimated accuracy of ± 10 days.

as an indication of the accuracy. According to this, there is not one chance in 10 that the true period differs from 11.04 years by more than 10 days. At first glance, this appears too good to be true.

But what does raw eyeballing of the data give? In 285 years, there are about $285/11 \approx 26$ cycles. If we can count these to an accuracy of $\pm 1/4$ cycle, our period estimate would be about

$$(f)_{\text{est}} = 11 \text{ years} \pm 39 \text{ days}.$$

Probability averaging of the noise, as discussed above (2.10), would reduce this uncertainty by about a factor of $\sqrt{285/10} = 5.3$, giving

$$(f)_{\text{est}} = 11 \text{ years} \pm 7.3 \text{ days}, \quad \text{or} \quad (f)_{\text{est}} = 11 \pm 0.02 \text{ years}$$

which corresponds nicely with the result of the probability analysis.

These results come from analyzing the data by a model which said there is nothing present but a single sinusoid plus noise. Probability theory, given this model, is obliged to consider everything in the data that cannot be fit to a single sinusoid to be noise. But a glance at the data shows clearly that there is more present than our model assumed: therefore, probability theory must estimate the noise to be quite large.

This suggests that we might do better by using a more realistic model which allows the "signal" to have more structure. Such a model can be fit to the data more accurately; therefore it will estimate the noise to be smaller. This should permit a still better period estimate!

But caution forces itself upon us; by adding more and more components to the model we can always fit the data more and more accurately; it is absurd to suppose that by mere proliferation of a model we can extract arbitrarily accurate estimates of a parameter. There must be a point of diminishing returns – or indeed of negative returns – beyond which we are deceiving ourselves.

It is very important to understand the following point. Probability theory always gives us the estimates that are justified by the information *that was actually used* in the calculation. Generally, a person who has more relevant information will be able to do a different (more complicated) calculation, leading to better estimates. But of course, this presupposes that the extra information is actually true. If one puts false information into a probability calculation, then probability theory will give optimal estimates based on false information: these could be very misleading. The onus is always on the user to tell the truth and nothing but the truth; probability theory has no safety device to detect falsehoods.

The issue just raised takes us into an area that has been heretofore, to the best of our knowledge, unexplored by any coherent theory. The analysis of this section has shown how to make the optimum estimates of parameters *given a model* whose correctness is not questioned. Deeper probability analysis is needed to indicate how to make the optimum choice of a model, which neither cheats us by giving poorer estimates than the data could justify, nor deceives us by seeming to give better estimates than the data can justify. But before we can turn to the model selection problem, the results of this chapter must be generalized to more complex models and it is to this task that we now turn.

Chapter 3

THE GENERAL MODEL EQUATION PLUS NOISE

The results of the previous chapter already represent progress on the spectral analysis problem because we were able to remove consideration of the amplitude, phase and noise level, and find what probability theory has to say about the frequency alone. In addition, it has given us an indication about how to proceed to more general problems. If we had used a model where the quadratic term in the likelihood function did not simplify, we would have a more complicated analytical solution. Although any multivariate Gaussian integral can be done, the key to being able to remove the nuisance parameters easily, and above all selectively, was that the likelihood factored into independent parts. In the full spectrum analysis problem worked on by Jaynes, [12] the nuisance parameters were not independent, and the explicit solution required the diagonalization of a matrix that could be quite large.

3.1 The Likelihood Function

To understand an easier approach to complex models, suppose we have a model of the form

$$d_i = f(t_i) + e_i$$

$$f(t) = \sum_{j=1}^{m} B_j G_j(t, \{\omega\}). \tag{3.1}$$

The model functions, $G_i(t, \{\omega\})$, are themselves functions of a set of parameters which we label collectively $\{\omega\}$ (these parameters migh', be frequencies, chirp rates, decay rates, or any other quantities one could encounter). Now if we substitute this model into the likelihood (2.2), the simplification that occurred in (2.3) does not take place:

$$L(\{B\},\{\omega\},\sigma) \propto \sigma^{-N} \times \exp\{-\frac{NQ}{2\sigma^2}\} \tag{3.2}$$

where

$$Q \equiv \overline{d^2} - \frac{2}{N}\sum_{j=1}^{m}\sum_{i=1}^{N} B_j d_i G_j(t_i) + \frac{1}{N}\sum_{j=1}^{m}\sum_{k=1}^{m} g_{jk} B_j B_k \tag{3.3}$$

$$g_{jk} = \sum_{i=1}^{N} G_j(t_i) G_k(t_i). \tag{3.4}$$

If the desired simplification is to take place, the matrix g_{jk} must be diagonal.

3.2 The Orthonormal Model Equations

For the matrix g_{jk} to be diagonal the model functions G_j must be made orthogonal. This can be done by taking appropriate linear combinations of them. But care must be taken; we do not desire a set of orthogonal functions of a continuous variable t, but a set of vectors which are orthogonal when summed over the discrete sampling times t_i. It is the sum over t_i appearing in the quadratic term of the likelihood which must simplify.

To accomplish this, consider the real symmetric matrix g_{jk} (3.4) defined above. Since for all x_j satisfying $\sum x_j^2 > 0$,

$$\sum_{j,k=1}^{m} g_{jk} x_j x_k = \sum_{i=1}^{N}\left(\sum_{j=1}^{m} x_j G_j(t_i)\right)^2 \geq 0$$

so that g_{jk} is positive definite if it is of rank m. If it is of rank $r < m$, then the model functions $G_j(t)$ or the sampling times t_i were poorly chosen. That is, if a linear combination of the $G_j(t)$ is zero at every sampling point:

$$\sum_{j=1}^{m} x_j G_j(t_i) = 0, \qquad (1 \leq i \leq N)$$

then at least one of the model functions $G_j(t)$ is redundant and can be removed from the model without changing the problem.

We suppose that redundant model functions have been removed, so that g_{jk} is positive definite and of rank m in what follows. Let e_{kj} represent the jth component of the kth normalized eigenvector of g_{jk}; i.e.

$$\sum_{k=1}^{m} g_{jk}e_{lk} = \lambda_l e_{lj},$$

where λ_l is the lth eigenvalue of g_{jk}. Then the functions $H_j(t)$, defined as

$$H_j(t) = \frac{1}{\sqrt{\lambda_j}} \sum_{k=1}^{m} e_{jk}G_k(t), \tag{3.5}$$

have the desired orthonormality condition,

$$\sum_{i=1}^{N} H_j(t_i)H_k(t_i) = \delta_{jk}. \tag{3.6}$$

The model Eq. (3.1) can now be rewritten in terms of these orthonormal functions as

$$f(t) = \sum_{k=1}^{m} A_k H_k(t). \tag{3.7}$$

The amplitudes B_k are linearly related to the A_k by

$$B_k = \sum_{j=1}^{m} \frac{A_j e_{jk}}{\sqrt{\lambda_j}} \quad \text{and} \quad A_k = \sqrt{\lambda_k} \sum_{j=1}^{m} B_j e_{kj}. \tag{3.8}$$

The volume elements are given by

$$\begin{aligned} dB_1 \cdots dB_m d\omega_1 \cdots d\omega_r &= \left| \frac{e_{lj}}{\sqrt{\lambda_j}} \right| dA_1 \cdots dA_m d\omega_1 \cdots d\omega_r \\ &= \lambda_1^{-\frac{1}{2}} \cdots \lambda_m^{-\frac{1}{2}} dA_1 \cdots dA_m d\omega_1 \cdots d\omega_r. \end{aligned} \tag{3.9}$$

The Jacobian is a function of the $\{\omega\}$ parameters and is a constant as long as we are not integrating over these $\{\omega\}$ parameters. At the end of the calculation the linear relations between the A's and B's can be used to calculate the expected values of the B's from the expected value of the A's, and the same is true of the second posterior moments:

$$E(B_k|\{\omega\}, D, I) = \langle B_k \rangle = \sum_{j=1}^{m} \frac{\langle A_j \rangle e_{jk}}{\sqrt{\lambda_j}} \tag{3.10}$$

$$E(B_k B_l|\{\omega\}, D, I) = \langle B_k B_l \rangle = \sum_{i=1}^{m} \sum_{j=1}^{m} \frac{e_{ik} e_{jl} \langle A_i A_j \rangle}{\sqrt{\lambda_i \lambda_j}} \tag{3.11}$$

where $E(B_k|D, I)$ stands for the expectation value of B_k given the data D, and the prior information I: this is the notation used by the general statistical community, while $\langle B_k \rangle$ is the notation more familiar in the physical sciences.

The two operations of making a transformation on the model functions and changing variables will transform any nonorthonormal model of the form (3.1) into an orthonormal model (3.7). We still have a matrix to diagonalize, but this is done once at the beginning of the calculation. If the g_{jk} matrix cannot be diagonalized analytically, it can still be computed numerically and then diagonalized. It is not necessary to carry out the inverse transformation if we are interested only in estimating the $\{\omega\}$ parameters: the $H_j(t, \{\omega\})$ are functions of them.

3.3 Elimination of the Nuisance Parameters

We are now in a position to proceed as before. Because the calculation is essentially identical to the single harmonic frequency calculation we will proceed very rapidly. The likelihood can now be factored into a set of independent likelihoods for each of the A_j. It is now possible to remove the nuisance parameters easily. Using the joint likelihood (3.2), we make the change of function (3.5) and the change of variables (3.8) to obtain the joint likelihood of the new parameters

$$L(\{A\}, \{\omega\}, \sigma) \propto \sigma^{-N} \times \exp\left\{-\frac{N}{2\sigma^2}[\overline{d^2} - \frac{2}{N}\sum_{j=1}^{m} A_j h_j + \frac{1}{N}\sum_{j=1}^{m} A_j^2]\right\} \tag{3.12}$$

$$h_j \equiv \sum_{i=1}^{N} d_i H_j(t_i), \qquad (1 \le j \le m). \tag{3.13}$$

Here h_j is just the projection of the data onto the orthonormal model function H_j. In the simple harmonic analysis performed in Chapter 2, the $R(\omega)$ and $I(\omega)$ functions are the analogues of these h_j functions. However, the h_j functions are more general, we did not make any approximations in deriving them. The orthonormality of the H_j functions was used to simplify the quadratic term. This simplification makes it possible to complete the square in the likelihood and to integrate over the A_j's, or any selected subset of them.

As before, if one has prior information about these amplitudes, then here is where it should be incorporated. Because we are performing a general calculation and have not specified the model functions we assume no prior information is available about the amplitudes, and thus obtain conservative estimates by assigning the amplitudes a uniform prior. Performing the m integrations one obtains

$$L(\{\omega\}, \sigma) \propto \sigma^{-N+m} \times \exp\left\{-\frac{N\overline{d^2} - m\overline{h^2}}{2\sigma^2}\right\} \tag{3.14}$$

where

$$\overline{h^2} \equiv \frac{1}{m}\sum_{j=1}^{m} h_j^2 \tag{3.15}$$

is the mean-square of the observed projections. This equation is the analogue of (2.6) in the simple harmonic calculation. Although it is exact and far more general, it is actually simpler in structure and gives us a better intuitive understanding of the problem than (2.6), as we will see in the Bessel inequality below. In a sense $\overline{h^2}$ is a generalization of the periodogram to arbitrary model functions. In its dependence on the parameters $\{\omega\}$ it is a sufficient statistic for all of them.

If σ is known, then the problem is again completed, provided we have no additional prior information. The joint posterior probability of the $\{\omega\}$ parameters, conditional on the data and our knowledge of σ, is

$$P(\{\omega\}|D,\sigma,I) \propto \exp\left\{\frac{m\overline{h^2}}{2\sigma^2}\right\}. \tag{3.16}$$

But if there is no prior information available about the noise, then σ is a nuisance parameter and can be eliminated as before. Using the Jeffreys prior $1/\sigma$ and integrating (3.14) over σ gives

$$P(\{\omega\}|D,I) \propto \left[1 - \frac{m\overline{h^2}}{N\overline{d^2}}\right]^{\frac{m-N}{2}}. \tag{3.17}$$

This is again of the general form of the "Student t-distribution" that we found before in (2.8). But one may be troubled by the negative sign [in the square brackets (3.17)], which suggests that (3.17) might become singular. We pause to investigate this possibility by Bessel's famous argument.

3.4 The Bessel Inequality

Suppose we wish to approximate the data vector $\{d_1, \cdots, d_N\}$ by the orthogonal functions $H_j(t)$:

$$d_i = \sum_{j=1}^{m} a_j H_j(t_i) + \text{error}, \qquad (1 \le i \le N).$$

What choice of $\{a_1, \cdots, a_m\}$ is "best"? If our criterion of "best" is the mean-square error, we have

$$\begin{aligned} 0 &\leq \sum_{i=1}^{N} \left[d_i - \sum_{j=1}^{m} a_j H_j(t_i) \right]^2 \\ &= N\overline{d^2} + \sum_{j=1}^{m} (a_j^2 - 2a_j h_j) \\ &= N\overline{d^2} - m\overline{h^2} + \sum_{j=1}^{m} (a_j - h_j)^2 \end{aligned}$$

where we have used (3.13) and the orthonormality (3.6). Evidently, the "best" choice of the coefficients is

$$a_j = h_j, \qquad (1 \leq j \leq m)$$

and with this choice the minimum possible mean-square error is given by the Bessel inequality

$$N\overline{d^2} - m\overline{h^2} \geq 0 \tag{3.18}$$

with equality if and only if the approximation is perfect. In other words, Eq. (3.17) becomes singular somewhere in the parameter space if and only if the model

$$f(t) = \sum_{j=1}^{m} A_j H_j(t)$$

can be fitted to the data exactly. But in that case we know the parameters by deductive reasoning, and probability theory becomes superfluous. Even so, probability theory is still working correctly, indicating an infinitely greater probability for the true parameter values than for any others.

3.5 An Intuitive Picture

The Bessel inequality gives us the following intuitive picture of the meaning of Eqs. (3.12-3.17). The data $\{d_1, \cdots, d_N\}$ comprise a vector in an N-dimensional linear vector space S_N. The model equation

$$d_i = \sum_{j=1}^{m} A_j H_j(t_i) + e_i, \qquad (1 \leq i \leq N)$$

supposes that these data can be separated into a "systematic part" $f(t_i)$ and a "random part" e_i. Estimating the parameters of interest $\{\omega\}$ that are hidden in the model functions $H_j(t)$ amounts essentially to finding the values of the $\{\omega\}$ that permit $f(t)$ to make the closest possible fit to the data by the mean-square criterion. Put differently, probability theory tells us that the most likely values of the $\{\omega\}$ are those that allow a maximum amount of the mean-square data $\overline{d^2}$ to be accounted for by the systematic term; from (3.18), those are the values that maximize $\overline{h^2}$.

However, we have N data points and only m model functions to fit to them. Therefore, to assign a particular model is equivalent to supposing that the systematic component of the data lies only in an m-dimensional subspace S_m of S_N. What kind of data should we then expect?

Let us look at the problem backwards for a moment. Suppose someone knows (never mind how he could know this) that the model is correct, and he also knows the true values of all the model parameters $(\{A\}, \{\omega\}, \sigma)$ – call this the Utopian state of knowledge U – but he does not know what data will be found. Then the probability density that he would assign to any particular data set $D = \{d_1, \cdots, d_N\}$ is just our original sampling distribution (3.2):

$$P(D|U) = (2\pi\sigma^2)^{-\frac{N}{2}} \exp\left\{-\frac{1}{2\sigma^2}\sum_{i=1}^{N}[d_i - f(t_i)]^2\right\}.$$

From this he would find the expectations and covariances of the data:

$$\begin{aligned} E(d_i|U) = \langle d_i \rangle &= f(t_i) \qquad (1 \le i \le N) \\ \langle d_i d_j \rangle - \langle d_i \rangle \langle d_j \rangle &= (2\pi\sigma^2)^{-\frac{N}{2}} \int d^N x \; x_i x_j \exp\left\{-\frac{1}{2\sigma^2}\sum_{i=1}^{N} x_i^2\right\} \\ &= \sigma^2 \delta_{ij} \end{aligned}$$

therefore he would "expect" to see a value of $\overline{d^2}$ of about

$$\begin{aligned} E(\overline{d^2}|U) = \langle \overline{d^2} \rangle &= \frac{1}{N}\sum_{i=1}^{N}\langle d_i^2 \rangle \\ &= \frac{1}{N}\sum_{i=1}^{N}(\langle d_i \rangle^2 + \sigma^2) \\ &= \frac{1}{N}\sum_{i=1}^{N} f^2(t_i) + \sigma^2. \end{aligned} \tag{3.19}$$

But from the orthonormality (3.6) of the $H_j(t_i)$, we have

$$\begin{aligned} \sum_{i=1}^{N} f^2(t_i) &= \sum_{l=1}^{N} \sum_{j,k=1}^{m} A_j A_k H_j(t_i) H_k(t_i) \\ &= \sum_{j=1}^{m} A_j^2. \end{aligned}$$

So that (3.19) becomes

$$\langle\overline{d^2}\rangle = \frac{m}{N}\overline{A^2} + \sigma^2.$$

Now, what value of $\overline{h^2}$ would he expect the data to generate? This is

$$\begin{aligned} E(\overline{h^2}|U) = \langle\overline{h^2}\rangle &= \frac{1}{m}\sum_{j=1}^{m}\langle h_j^2\rangle \\ &= \frac{1}{m}\sum_{j=1}^{m}\sum_{i,k=1}^{N}\langle d_i d_k\rangle H_j(t_i)H_j(t_k) \qquad (3.20) \\ &= \frac{1}{m}\sum_{j=1}^{m}\sum_{i,k=1}^{N}(\langle d_i\rangle\langle d_k\rangle + \sigma^2\delta_{ik})H_j(t_i)H_j(t_k). \end{aligned}$$

But

$$\begin{aligned} \sum_{i=1}^{N}\langle d_i\rangle H_j(t_i) &= \sum_{i=1}^{N}\sum_{l=1}^{m} A_l H_l(t_i)H_j(t_i) \\ &= \sum_{l=1}^{m} A_l\delta_{lj} \\ &= A_j \end{aligned}$$

and (3.20) reduces to

$$\langle\overline{h^2}\rangle = \overline{A^2} + \sigma^2.$$

So he expects the left-hand side of the Bessel inequality (3.18) to be approximately

$$N\langle\overline{d^2}\rangle - m\overline{h^2} \approx (N-m)\sigma^2. \qquad (3.21)$$

This agrees very nicely with our intuitive judgment that as the number of model functions increases, we should be able to fit the data better and better. Indeed, when $m = N$, the $H_j(t_i)$ become a complete orthonormal set on S_N, and the data can always be fit exactly, as (3.21) suggests.

3.6 A Simple Diagnostic Test

If σ is known, these results give a simple diagnostic test for judging the adequacy of our model. Having taken the data, calculate $(N\overline{d^2} - m\overline{h^2})$. If the result is reasonably close to $(N-m)\sigma^2$, then the validity of the model is "confirmed" (in the sense that the data give no evidence against the model). On the other hand, if $(N\overline{d^2} - m\overline{h^2})$

turns out to be much larger than $(N-m)\sigma^2$, the model is not fitting the data as well as it should: it is "underfitting" the data. That is evidence either that the model is inadequate to represent the data; we could need more model functions, or different model functions, or our supposed value of σ^2 is too low. The next order of business would be to investigate these possibilities.

It is also possible, although unusual, that $(N\overline{d^2} - m\overline{h^2})$ is far less than $(N-m)\sigma^2$; the model is "overfitting" the data. That is evidence either that our supposed value of σ is too large (the data are actually better than we expected), or that the model is more complex than it needs to be. By adding more model functions we can always improve the apparent fit, but if our model functions represent more detail than is really in the systematic effects at work, part of this fit is misleading: we are *fitting the noise.*

A test to confirm this would be to repeat the whole experiment under conditions where we know the parameters should have the same values as before, and compare the parameter estimates from the two experiments. Those parameters that are estimated to be about the same in the two experiments are probably real systematic effects. If some parameters are estimated to be quite different in the two experiments, they are almost surely spurious: i.e. these are not real effects but only artifacts of fitting the noise. The model should then be simplified, by removing the spurious parameters.

Unfortunately, a repetition is seldom possible with geophysical or economic time series, although one may split the data into two parts and see if they make about the same estimates. But repetition is usually easy and standard practice in the controlled environment of a physics experiment. Indeed, the physicist's common-sense criterion of a real effect is its reproducibility. Probability theory does not conflict with good common-sense judgment; it only sharpens it and makes it quantitative. A striking example of this is given in the scenario below.

Consider now the case that σ is completely unknown, where probability theory led us to (3.17). As we show in Appendix C, integrating over a nuisance parameter is very much like estimating the parameter from the data, and then using that estimate in our equations. If the parameter is actually well determined by the data, the two procedures are essentially equivalent. In Chapter 4 we derive an exact expression for the expectation value of the variance $\langle\sigma^2\rangle$:

$$\begin{aligned}\langle\sigma^2\rangle &= \frac{N}{N-m-2}\left[\overline{d^2}-\frac{m\overline{h^2}}{N}\right]\\ &= \frac{1}{N-m-2}\left[\sum_{i=1}^{N} d_i^2-\sum_{j=1}^{m} h_j^2\right].\end{aligned} \tag{3.22}$$

Constraining σ to this value, we obtain for the posterior probability of the $\{\omega\}$ parameters approximately

$$P(\{\omega\}|D,\langle\sigma^2\rangle,I)\approx \exp\left\{\frac{m\overline{h^2}}{\langle\sigma^2\rangle}\right\}.$$

In effect, probability theory tells us that we should apportion the first m degrees of freedom to the signal, the next two to the variance, and the remaining $(N-m-2)$ should be noise degrees of freedom. Thus everything probability theory cannot fit to the signal will be placed in the noise.

More interesting is the opposite extreme when (3.17) approaches a singular value. Consider the following scenario. You have obtained some data which are recorded automatically to six figures and look like this: $D=\{d_1=1423.16, d_2=1509.77, d_3=1596.38,\cdots\}$. But you have no prior knowledge of the accuracy of those data; for all you know, σ may be as large as 100 or even larger, making the last four digits garbage. But you plot the data, to determine a model function that best fits them. Suppose, for simplicity, that the model function is linear: $d_i=a+si+e_i$. On plotting d_i against i, you are astonished and delighted to see the data falling exactly on a straight line (i.e. to within the six figures given). What conclusions do you draw from this?

Intuitively, one would think that the data must be far "better" than had been thought; you feel sure that $\sigma<10^{-2}$, and that you are therefore able to estimate the slope s to an accuracy considerably better than $\pm 10^{-2}$, if the number of data values N is large. It may, however, be hard to see at first glance how probability theory can justify this intuitive conclusion that we draw so easily.

But that is just what (3.17) and (3.22) tell us; Bayesian analysis leads us to it automatically and for any model functions. Even though you had no reason to expect it, if it turns out that the data can be fit almost exactly to a model function, then from the Bessel inequality (3.18) it follows that σ^2 must be extremely small and, if the other parameters are independent, they can all be estimated almost exactly.

By "independent" in the last paragraph we mean that a given model function $f(t)=\sum A_j H_j(t)$ can be achieved with only one unique set of values for the parameters. If several different choices of the parameters all lead to the same model function, of course the data cannot distinguish between them; only certain functions of the parameters can be estimated accurately, however many data we have. In this case the parameters are said to be "confounded" or "unidentified". Generally, this would be a sign that the model was poorly chosen. However, it may be that the parameters are known to be real, and the experiment, whether by poor design or the perversity of nature, is just not capable of distinguishing them.

As an example of confounded parameters, suppose that two different sinusoidal signals are known to be present, but they have identical frequencies. Then their separate amplitudes are confounded: the data can give evidence only about their sum. The difference in amplitudes can be known only from prior information.

Chapter 4

ESTIMATING THE PARAMETERS

Once the models had been rewritten in terms of the orthonormal model functions, we were able to remove the nuisance parameters $\{A\}$ and σ. The integrals performed in removing the nuisance parameters were all Gaussian or gamma integrals; therefore, one can always compute the posterior moments of these parameters.

There are a number of reasons why these moments are of interest: the first moments of the amplitudes are needed if one intends to reconstruct the model $f(t)$; the second moments are related to the energy carried by the signal; the estimated noise variance σ^2 and the energy carried by the signal can be used to estimate the signal-to-noise ratio of the data. Thus the parameters $\{A\}$ and σ are not entirely "nuisance" parameters; it is of some interest to estimate them. Additionally, we cannot in general compute the expected value of the $\{\omega\}$ parameters analytically. We must devise a procedure for estimating these parameters and their accuracy.

4.1 The Expected Amplitudes $\langle A_j \rangle$

To begin we will compute the expected amplitudes $\langle A_j \rangle$ in the case where the variance is assumed known. Now the likelihood (3.12) is a function of the $\{\omega\}$ parameters and to estimate the $\langle A_j \rangle$ independently of the $\{\omega\}$'s, we should integrate over these parameters. Because we have not specified the model functions, we cannot do this once and for all. But we can obtain the estimate $\langle A_j \rangle$ as functions of the $\{\omega\}$ parameters. This gives us what would be the "best" estimate of the amplitudes if we knew the $\{\omega\}$ parameters.

The expected amplitudes are given by

$$E(A_j|\{\omega\},\sigma,D,I) = \langle A_j\rangle = \frac{\int_{-\infty}^{+\infty} dA_1\cdots dA_m A_j L(\{A\},\{\omega\},\sigma)}{\int_{-\infty}^{+\infty} dA_1\cdots dA_m L(\{A\}|\{\omega\},\sigma)}.$$

We will carry out the first integration in detail to illustrate the procedure, and later just give results. Using the likelihood (3.12) and having no prior information about A_j, we assign a uniform prior, multiply by A_j and integrate over the $\{A\}$. Because the joint likelihood is a product of their independent likelihoods, all of the integrals except the one over A_j cancel:

$$\langle A_j\rangle = \frac{\int_{-\infty}^{+\infty} dA_j A_j \exp\left\{-\frac{1}{2\sigma^2}[A_j^2 - 2A_j h_j]\right\}}{\int_{-\infty}^{+\infty} dA_j \exp\left\{-\frac{1}{2\sigma^2}[A_j^2 - 2A_j h_j]\right\}}.$$

A simple change of variables $u_j = (A_j - h_j)/\sqrt{2\sigma^2}$ reduces the integrals to

$$\langle A_j\rangle = \frac{\int_{-\infty}^{+\infty} du_j \left\{\sqrt{2\sigma^2}u_j + h_j\right\}\exp\left\{-u_j^2\right\}}{\int_{-\infty}^{+\infty} du_j \exp\left\{-u_j^2\right\}}.$$

The first integral in the numerator is zero from symmetry and the second gives

$$\langle A_j\rangle = h_j. \tag{4.1}$$

This is the result one would expect. After all, we are expanding the data on an orthonormal set of vectors. The expansion coefficients are just the projections of the data onto the expansion vectors and that is what we find.

We can use these expected amplitudes $\langle A_j\rangle$ to calculate the expectation values of the amplitudes $\langle B_j\rangle$ in the nonorthogonal model. Using (3.10), these are given by

$$E(B_j|\{\omega\},\sigma,D,I) = \langle B_k\rangle = \sum_{j=1}^{m} \frac{h_j e_{jk}}{\sqrt{\lambda_j}}. \tag{4.2}$$

Care must be taken in using this formula, because the dependence of the $\langle B_k\rangle$ on the $\{\omega\}$ parameters is hidden. The functions h_j, the eigenvectors e_{kj} and the eigenvalues λ_j are all functions of the $\{\omega\}$ parameters. To remove the $\{\omega\}$ dependence from (4.2) one must multiply by $P(\{\omega\}|D,I)$ and integrate over all the $\{\omega\}$ parameters. If the total number of $\{\omega\}$ parameters r is large this will not be possible. Fortunately, if the total amount of data is large $P(\{\omega\}|D,I)$ will be so nearly a delta function that we can estimate these parameters from the maximum of $P(\{\omega\}|D,I)$.

Next we compute $\langle A_j \rangle$ when the noise variance σ^2 is unknown to see if obtaining independent information about σ will affect these results. To do this we need the likelihood $L(\{A\}, \{\omega\})$; this is given by (3.12) after removing the variance σ^2 using a Jeffreys prior $1/\sigma$:

$$L(\{\omega\}, \{A\}) \propto \left[\overline{d^2} - \frac{m\overline{h^2}}{N} + \frac{1}{N}\sum_{i=1}^{m}(A_j - h_j)^2\right]^{-\frac{N}{2}}. \tag{4.3}$$

Using (4.3) and repeating the calculation for $\langle A_j \rangle$ one obtains the same result (4.1). Apparently it does not matter if we know the variance or not. We will make the same estimate of the amplitudes regardless. As with some of the other results discovered in this calculation, this is what one's intuition might have said; knowing σ affects the accuracy of the estimates but not their actual values. Indeed, the first moments were independent of the value of σ when the variance was known; it is hard to see how the first moments could suddenly become different when the variance is unknown.

4.2 The Second Posterior Moments $\langle A_j A_k \rangle$

The second posterior moments $\langle A_j A_k \rangle$ cannot be independent of the noise variance σ^2, for that is what limits the accuracy of our estimates of the A_j. The second posterior moments, when the variance is assumed known, are given by

$$E(A_j A_k | \{\omega\}, \sigma, D, I) = \langle A_j A_k \rangle = \frac{\int_{-\infty}^{+\infty} dA_1 \cdots dA_m A_j A_k L(\{A\}, \{\omega\}, \sigma)}{\int_{-\infty}^{+\infty} dA_1 \cdots dA_m L(\{A\}, \{\omega\}, \sigma)}.$$

Using the likelihood (3.12) and again assuming a uniform prior, these expectation values are given by

$$\langle A_j A_k \rangle = h_j h_k + \sigma^2 \delta_{jk}$$

so that, in view of (4.1), the posterior covariances are

$$\langle A_j A_k \rangle - \langle A_j \rangle \langle A_k \rangle = \sigma^2 \delta_{jk}. \tag{4.4}$$

The A_j parameters are uncorrelated (we defined the model functions $H_j(t)$ to ensure this), and each one is estimated to an accuracy $\pm\sigma$. Intuitively, we might anticipate this but we would not feel very sure of it.

The expectation value $\langle A_j A_k \rangle$ may be related to the expectation value for the original model amplitudes by using (3.11):

$$\langle B_k B_l \rangle - \langle B_k \rangle \langle B_l \rangle = \sigma^2 \sum_{j=1}^{m} \frac{e_{jk} e_{jl}}{\lambda_j}. \tag{4.5}$$

These are the explicit Bayesian estimates for the posterior covariances for the original model. These are the most conservative estimates (in the sense discussed before) one can make, but they are still functions of the $\{\omega\}$ parameters.

We can repeat these calculations for the second posterior moments in the case when σ is assumed unknown to see if obtaining explicit information about σ is of use. Of course, we expect the results to differ from the previous result since (4.5) depends explicitly on σ. Performing the required calculation gives

$$\begin{aligned} E(A_jA_k|\{\omega\},D,I) &= \langle A_jA_k\rangle = h_jh_k \\ &+ \left[\frac{N}{N-2}\right]\left[\frac{2N-5}{2N-5-2m}\right]\left[\frac{2N-7}{2N-7-2m}\right]\left[\overline{d^2}-\frac{m\overline{h^2}}{N}\right]\delta_{jk}. \end{aligned}$$

Comparing this with (4.4) shows that obtaining independent information about σ will affect the estimates of the second moments. But not by much, as we will see below. The second term in this equation is essentially an estimate of σ^2, but for small N it differs appreciably from the direct estimate found next.

4.3 The Estimated Noise Variance $\langle\sigma^2\rangle$

One of the things that is of interest in an experiment is to estimate the noise power σ^2. This indicates how "good" the data appear to be in the light of the model, and can help one in making many judgments, such as whether to try a new model or build a new apparatus. We can obtain the expected value of σ as a function of the $\{\omega\}$ parameters; however, we can just as easily obtain the posterior moments $\langle\sigma^s\rangle$ for any power s. Using (3.14), and the Jeffreys prior $1/\sigma$, we integrate:

$$E(\sigma^s|\{\omega\},D,I) = \langle\sigma^s\rangle = \frac{\int_0^{+\infty} d\sigma\sigma^{s-1}L(\sigma|\{\omega\},D,I)}{\int_0^{+\infty} d\sigma\sigma^{-1}L(\sigma|\{\omega\},D,I)}$$

to obtain

$$\langle\sigma^s\rangle = \Gamma\left(\frac{N-m-s}{2}\right)\Gamma\left(\frac{N-m}{2}\right)^{-1}\left[\frac{N\overline{d^2}-m\overline{h^2}}{2}\right]^{\frac{s}{2}}. \tag{4.6}$$

For $s = 2$ this gives the estimated variance as

$$\begin{aligned} \langle\sigma^2\rangle &= \frac{N}{N-m-2}\left[\overline{d^2}-\frac{m\overline{h^2}}{N}\right] \\ &= \frac{1}{N-m-2}\left[\sum_{i=1}^{N} d_i^2 - \sum_{j=1}^{m} h_j^2\right]. \end{aligned} \tag{4.7}$$

The estimate depends on the number m of expansion functions used in the model. The more model functions we use, the smaller the last factor in (4.7), because by the Bessel inequality (3.18) the larger models fit the data better and $(\overline{d^2} - mN^{-1}\overline{h^2})$ decreases. But this should not decrease our estimate of σ^2 unless that factor decreases by more than we would expect from fitting the noise. The factor $N/(N-m-2)$ takes this into account. In effect probability theory tells us that $m+2$ degrees of freedom should go to estimating the model parameters and the variance, and the remaining degrees of freedom should go to the noise: everything not explicitly accounted for in the model is noise. We will show shortly that the estimated accuracy of the $\{\omega\}$ parameters depends directly on the estimated variance. If the model does not fit the data well, the estimates will become less precise in direct relation to the estimated variance.

We can use (4.6) to obtain an indication of the accuracy of the expected noise variance. The (mean) $\pm$ (standard deviation) estimate of σ^2 is

$$(\sigma^2)_{\text{est}} = \langle\sigma^2\rangle \pm \sqrt{\langle\sigma^4\rangle - \langle\sigma^2\rangle^2}.$$

From which we obtain

$$(\sigma^2)_{\text{est}} = \frac{N}{N-m-2}\left[\overline{d^2} - \frac{m\overline{h^2}}{N}\right]\left(1 \pm \epsilon\right)$$

$$\epsilon \equiv \sqrt{2/(N-m-4)}.$$

We then find the values of $N-m$ needed to achieve a given accuracy

% accuracy	ϵ	$N-m$
1	0.01	20,004
3	0.03	2,226
10	0.10	204
20	0.20	54

These are about what one would expect from simpler statistical estimation rules (the usual $N^{-\frac{1}{2}}$ rule of thumb).

4.4 The Signal-To-Noise Ratio

These results may be used to empirically estimate the signal-to-noise ratio of the data. We define this as the square root of the mean power carried by the signal divided by the mean power carried by the noise:

$$\frac{\text{Signal}}{\text{Noise}} = \left[(\sum_{j=1}^{m} A_j^2)/N\sigma^2 \right]^{\frac{1}{2}}.$$

This may be obtained from (4.2):

$$\frac{\text{Signal}}{\text{Noise}} = \left\{ \frac{m}{N} \left[1 + \frac{\overline{h^2}}{\sigma^2} \right] \right\}^{\frac{1}{2}}. \tag{4.8}$$

A similar empirical signal-to-noise ratio may be obtained when the noise variance σ is unknown by replacing σ in (4.8) by the estimated noise variance (4.7). When the data fit the model so well that $\overline{h^2} \gg \sigma^2$, the estimate reduces to

$$\left\{ \frac{m\overline{h^2}}{N\sigma^2} \right\}^{\frac{1}{2}} \quad \text{or} \quad \left\{ \frac{\sum_{j=1}^{m} h_j^2}{\sum_{k=1}^{N} e_i^2} \right\}^{\frac{1}{2}}$$

We will compute the signal-to-noise ratio for several models in the following sections.

4.5 Estimating the $\{\omega\}$ Parameters

Unlike the amplitudes $\{A\}$ and the variance σ^2, we cannot calculate the expectation values of the $\{\omega\}$ parameters analytically. In general, the integrals represented by

$$\langle \omega_j \rangle = \int d\omega_1 \cdots d\omega_r \, \omega_j \, P(\{\omega\}|D, I)$$

cannot be done exactly. Nonetheless we must obtain an estimate of these parameters, and their probable accuracy.

The exact joint posterior density is (3.16) when σ is known, and (3.17) when it is not. But they are not very different provided *we have enough data for good estimates.* For, writing the maximum attainable $\sum h_j^2$ as

$$\left(\sum_{j=1}^{m} h_j^2 \right)_{\max} = x$$

and writing the difference from the maximum as q^2 i.e.

$$\sum_{j=1}^{m} h_j^2 = x - q^2,$$

Eq. (3.17) becomes

$$\left[\sum_{i=1}^{N} d_i^2 - x + q^2 \right]^{\frac{m-N}{2}} \approx \exp\left\{ -\frac{(N-m)q^2}{2(\sum_{j=1}^{N} d_i^2 - x)} \right\}.$$

But this is nearly the same as

$$\left[\sum_{i=1}^{N} d_i^2 - x + q^2\right]^{\frac{m-N}{2}} \approx \exp\left\{-\frac{q^2}{2\langle\sigma^2\rangle}\right\}$$

where we used the estimate (4.7) for σ^2 evaluated for the values $\{\hat{\omega}\}$ that maximize the posterior probability as a function of the $\{\omega\}$ parameters. So up to an irrelevant normalization constant the posterior probability of the $\{\omega\}$ parameters around the location of the maximum is given by

$$P(\{\omega\}|D,\langle\sigma^2\rangle,I) \approx \exp\left\{\frac{m\overline{h^2}}{2\langle\sigma^2\rangle}\right\} \tag{4.9}$$

where the slightly inconsistent notation $P(\{\omega\}|\langle\sigma^2\rangle,D,I)$ has been adopted to remind us that we have used $\langle\sigma^2\rangle$, not σ^2. We have noted before that when we integrate over a nuisance parameter, the effect is for most purposes to estimate the parameter from the data, and then constrain the parameter to that value.

We expand $\overline{h^2}$, to obtain q^2, in a Taylor series around the maximum $\{\hat{\omega}\}$ to obtain

$$P(\{\omega\}|D,\langle\sigma^2\rangle,I) \propto \exp\left\{-\sum_{jk=1}^{r} \frac{b_{jk}}{2\langle\sigma^2\rangle}\Delta_j\Delta_k\right\} \tag{4.10}$$

where b_{jk} is the analogue of (2.9) defined in the single harmonic frequency problem

$$b_{jk} \equiv -\frac{m}{2}\frac{\partial^2\overline{h^2}}{\partial\omega_j\partial\omega_k} \tag{4.11}$$

$$\Delta_j \equiv \hat{\omega}_j - \omega_j.$$

From (4.10) we can make the (mean) ± (standard deviation) approximations for the $\{\omega\}$ parameters. We do these Gaussian integrals by first changing to orthogonal variables and then perform the r integrals just as we did with the amplitudes in Chapter 3. The new variables are obtained from the eigenvalues and eigenvectors of b_{jk}. Let u_{jk} denote the kth component of the jth eigenvector of b_{jk} and let v_j be the eigenvalue. The orthogonal variables are given by

$$s_j = \sqrt{v_j}\sum_{k=1}^{r}\Delta_k u_{kj} \qquad \Delta_j = \sum_{k=1}^{r}\frac{s_k u_{jk}}{\sqrt{v_k}}.$$

Making this change of variables, we have

$$P(\{s\}|\langle\sigma^2\rangle,D,I) \propto v_1^{-\frac{1}{2}}\cdots v_r^{-\frac{1}{2}}\exp\left\{-\sum_{j=1}^{r}\frac{s_j^2}{2\langle\sigma^2\rangle}\right\}. \tag{4.12}$$

From (4.12) we can compute $\langle s_j \rangle$ and $\langle s_j^2 \rangle$. Of course $\langle s_j \rangle$ is zero and the expectation value $\langle s_j s_k \rangle$ is given by

$$\langle s_k s_j \rangle = \frac{\int_{-\infty}^{\infty} ds_1 \cdots ds_r v_1^{-\frac{1}{2}} \cdots v_r^{-\frac{1}{2}} s_k s_j \exp\left\{-\sum_{l=1}^{r} \frac{s_l^2}{2\langle\sigma^2\rangle}\right\}}{\int_{-\infty}^{\infty} ds_1 \cdots ds_r v_1^{-\frac{1}{2}} \cdots v_r^{-\frac{1}{2}} \exp\left\{-\sum_{l=1}^{r} \frac{s_l^2}{2\langle\sigma^2\rangle}\right\}}$$

$$\langle s_k s_j \rangle = \langle \sigma^2 \rangle \delta_{kj}$$

where δ_{kj} is a Kronecker delta function. In the posterior distribution the s_j are uncorrelated, as they should be. From this we may obtain the posterior covariances of the $\{\omega\}$ parameters. These are

$$\langle \omega_j \omega_k \rangle - \langle \omega_j \rangle \langle \omega_k \rangle = \langle \sigma^2 \rangle \sum_{l=1}^{r} \frac{u_{lj} u_{lk}}{v_l},$$

and the variance γ_k^2 of the posterior distribution for ω_k is

$$\gamma_k^2 \equiv \langle \sigma^2 \rangle \sum_{j=1}^{r} \frac{u_{jk}^2}{v_j}. \tag{4.13}$$

Then the estimated ω_j parameters are

$$(\omega_j)_{\text{est}} = \hat{\omega}_j \pm \gamma_j \tag{4.14}$$

and; here $\hat{\omega}_j$ is the location of the maximum of the probability distribution as a function of the $\{\omega\}$ parameter.

For an arbitrary model the matrix b_{jk} cannot be calculated analytically; however, it can be evaluated numerically using the computer code given in Appendix E. We use a general searching routine to find the maximum of the probability distribution and then calculate this matrix numerically. The log of the "Student t-distribution" is so sharply peaked that gradient searching routines do not work well. We use a "pattern" search routine described by Hooke and Jeeves [21] [22].

The accuracy estimates of both the $\{\omega\}$ parameters and the amplitudes $\{A\}$ in Eq. (4.5) depend explicitly on the estimated noise variance. But the estimated variance is the mean square difference between the model and the data. If the misfit is large the variance is estimated to be large and the accuracy is estimated to be poor. Thus when we say that the parameter estimates are conservative we mean that, because everything probability theory cannot fit to the model is assigned to the noise, all of our parameter estimates are as wide as is consistent with the model and the data. For example, when we estimate a frequency from a discrete Fourier transform we are in effect using a single harmonic frequency model for an estimate (position of a peak). But the width of the peak has nothing to do with the noise level, and if we supposed it, erroneously, to be an indication of the accuracy of our estimate, we could make very large errors.

This is perhaps one of the most subtle and important points about the use of uninformative priors that comes out in this work, and we will try to state it more clearly. When we did this calculation, at every point where we had to supply a prior probability we chose a prior that was as uninformative as possible (by uninformative we mean that the prior is as smooth as it can be and still be consistent with the known information). Specifically we mean a prior that has no sharp maximum: one that does not determine any value of the parameter strongly. We derived the Gaussian for the noise prior as the smoothest, least informative, prior that was consistent with the given second moment of the noise. We specifically did not assume the noise was nonwhite or correlated because we do not have prior information to that effect. So if the noise turns out to be colored we have in effect already allowed for that possibility because we used a less informative prior for the noise, which automatically considers every possible way of being colored, in the sense that the white noise basic support set includes all those of colored noise. On the other hand, if we knew a specific way in which the noise departs from whiteness, we could exploit that information to obtain a more concentrated noise probability distribution, leading to still better estimates of the $\{\omega\}$ parameters. We will demonstrate this point several times in Chapter 6.

4.6 The Power Spectral Density

Although not explicitly stated, we have calculated above an estimate of the total energy of the signal. The total energy E carried by the signal in our orthogonal model is

$$E \equiv \int_{t_1}^{t_N} f(t)^2 dt \approx \sum_{j=1}^{m} A_j^2$$

and its spectral density is given by

$$\hat{p}(\{\omega\}) = m\left[\sigma^2 + \overline{h^2}\right] P(\{\omega\}|D, I, \sigma). \tag{4.15}$$

This function is the energy per unit $\{\omega\}$ carried by the signal (not the noise). This power spectral estimate is essentially a power normalized probability distribution, and should not be confused with what a power meter would measure (which is the total power carried by the signal and the noise).

We have seen this estimated variance term once before. When we derived the power spectral density for the single harmonic frequency a similar term was present [see Eq. (2.16)]. That term of $m\sigma^2$ in (4.15) might be a little disconcerting to some; if (4.15) estimates the energy carried by the "signal" why does it include the noise power σ^2? If $\overline{h^2} \gg \sigma^2$ then the term is of no importance. But in the unlikely event $\overline{h^2} \ll \sigma^2$, then what is this term telling us? When these equations were formulated we put in the fact that there is present noise of variance σ^2 in a space of dimension N, and a signal in a subspace of m model functions. But then if $\overline{h^2} \ll \sigma^2$, there is only one explanation: the noise is such that its components on those m model functions just happened to cancel the signal. But if the noise just cancelled the signal, then the power carried by the signal must have been equal to the power $m\sigma^2$ carried by the noise in those m functions; and that is exactly the answer one obtains. This is an excellent example of the sophisticated subtlety of Bayesian analysis, which automatically perceives things that our unaided intuition might not (and indeed did not) notice in years of thinking about such problems.

We have already approximated $P(\{\omega\}|D,\sigma,I)$ as a Gaussian expanded about the maximum of the probability density. Using (4.10) we can approximate the power spectral density as

$$\begin{aligned} \hat{p}(\{\omega\}) &\approx m[\langle\sigma^2\rangle + \overline{h^2}]P(\{\omega\}|\langle\sigma^2\rangle, D, I) \\ P(\{\omega\}|\langle\sigma^2\rangle, D, I) &\propto \exp\left\{-\sum_{jk=1}^{r} \frac{b_{jk}(\hat{\omega}_j - \omega_j)(\hat{\omega}_k - \omega_k)}{2\langle\sigma^2\rangle}\right\}. \end{aligned} \tag{4.16}$$

This approximation will turn out to be very useful. We will be dealing typically with problems where the $\{\omega\}$ parameters are well determined or where we wish to remove one or more of the $\{\omega\}$ parameters as nuisances. For example, when we plot the power spectral density for multiple harmonic frequencies, we do not wish to plot this as a function of multiple variables, but as a function of one frequency: all other frequencies must be removed by integration. We cannot do these integrals in (4.15); in general, however, we will be able to do them in (4.16).

There are two possible problems with this definition of the power spectral density. First we assumed there is only one maximum in the posterior probability density, and second we asked a question about the total power carried by the signal, not a question about one spectral line. It will turn out that the multiple frequency model will be invariant under permutations of the labels on the frequencies. It cannot matter which frequency is number one and which is labeled number two. This invariance must manifest itself in the joint posterior probability density; there will be multiple peaks of equal probability and we will be led to generalize this definition. In addition we ask a question about the total energy carried by the signal in the sampling time. This is the proper question when the signal is not composed of sinusoids. But asking a question about the total energy is not the same as asking about the energy carried by each sinusoid. We will need to introduce another quantity that will describe the energy carried by one sinusoid. Before we do this we need to understand much more about the problem of estimating frequencies and decay rates. Chapter 6 is devoted primarily to this subject. For now we turn attention to a slightly more general problem of "how to make the optimal choice of a model?"

Chapter 5

MODEL SELECTION

When analyzing the results of an experiment it is not always known which model function (3.1) applies. We need a way to choose between several possible models. This is easily done using Bayes' theorem (1.3) and repeated applications of the procedure (1.4) which led to the "Student t-distribution." The first step in answering this question is to enumerate the possible models. Suppose we have a set of s possible models $\{H_1, \cdots, H_s\}$ with model functions $\{f_1, \cdots, f_s\}$. We are hardly ever sure that the "true" model is actually contained in this set. Indeed, the "set of all possible models" is not only infinite, but it is also quite undefined. It is not even clear what one could mean by a "true" model; both questions may take us into an area more like theology than science.

The only questions we seek to examine here are the ones that are answerable because they are mathematically well-posed. Such questions are of the form: "Given a specified set S_s of possible models $\{H_1, \cdots, H_s\}$ and looking only within that set, which model is most probable in view of all the data and prior information, and how strongly is it supported relative to the alternatives in that set?" Bayesian analysis can give a definite answer to such a question – see [15], [23].

5.1 What About "Something Else?"

To say that we confine ourselves to the set S_s is not to assert dogmatically that there are no other possibilities; we may assign prior probabilities $P(H_j|I), (1 \leq j \leq s)$

which do not add up to one:

$$\sum_{j=1}^{s} P(H_j|I) = a < 1.$$

Then we are assigning a prior probability $(1 - a)$ to some unknown proposition

$$\text{SE} \equiv \text{“Something Else not yet thought of.”}$$

But until SE is specified it cannot enter into a Bayesian analysis; probability theory can only compare the specified models $\{f_1, \cdots, f_s\}$ with each other.

Let us demonstrate this more explicitly. If we try to include SE in our set of hypotheses, we can calculate the posterior probabilities of the $\{f_j\}$ and SE to obtain

$$P(f_j|D, I) = \frac{P(f_j|I)P(D|f_j, I)}{P(D|I)}$$

and

$$P(\text{SE}|D, I) = \frac{P(\text{SE}|I)P(D|\text{SE}, I)}{P(D|I)}.$$

But this is numerically indeterminate even if $P(\text{SE}|I) = 1 - a$ is known, because $P(D|\text{SE}, I)$ is undefined until that "Something Else" is specified. The denominator $P(D|I)$ is also indeterminate, because

$$\begin{aligned} P(D|I) &= \sum_{j=1}^{s} P(D, f_j|I) + P(D, \text{SE}|I) \\ &= \sum_{j=1}^{s} P(D|f_j, I)P(f_j|I) + P(D|\text{SE}, I)P(\text{SE}|I). \end{aligned}$$

But the relative probabilities of the specified models are still well defined, because the indeterminates cancel out:

$$\frac{P(f_i|D, I)}{P(f_j|D, I)} = \frac{P(f_i|I)}{P(f_j|I)} \frac{P(D|f_i, I)}{P(D|f_j, I)}.$$

These relative probabilities are independent of what probability $(1 - a)$ we assign to "Something Else", so we shall get the same results if we just ignore "Something Else" altogether, and act as if $a = 1$. In other words, while it is not wrong to introduce an unspecified "Something Else" into a probability calculation, no useful purpose is served by it, and we shall not do so here.

5.2 The Relative Probability of Model f_j

We wish to confine our attention to a selected set of models $\{f_1, \cdots, f_s\}$. Because of the arguments just given we may write

$$\sum_{j=1}^{s} P(f_j|D,I) = 1$$

where $P(f_j|D,I)$ is the posterior probability of model f_j. From Bayes' theorem (1.3) we may write

$$P(f_j|D,I) = \frac{P(f_j|I)P(D|f_j,I)}{P(D|I)} \tag{5.1}$$

and

$$P(D|I) = \sum_{j=1}^{s} P(f_j|I)P(D|f_j,I).$$

The way to proceed on this problem is to apply the general procedure for removing nuisance parameters given in Chapter 1. We will assume for now that the variance of the noise σ^2 is known and derive $P(f_j|\sigma,D,I)$, then at the end of the calculation if σ is not known we will remove it. Thus symbolically, we have

$$P(D|\sigma,f_j,I) = \int d\{A\}d\{\omega\}P(\{A\},\{\omega\}|I)P(D|\{A\},\{\omega\},\sigma,f_j,I). \tag{5.2}$$

But this is essentially just the problem we solved in Chapter 3 [Eqs. (3.12-3.17)] with three additions: when there can be differing numbers of parameters we must use normalized priors, we must do the integrals over the $\{\omega\}$ parameters, and the direct probability Eq. (5.2) of the data for the jth model must include all numerical factors in $P(D|\{A\},\{\omega\},\sigma,f_j,I)$. We will need to keep track of the normalization constants explicitly because the results we obtain will depend on them. We will do this calculation in four steps; first perform the integrals over the amplitudes $\{A\}$ using an appropriately normalized prior. Second we approximate the quasi-likelihood of the $\{\omega\}$ parameters about the maximum likelihood point; third remove the $\{\omega\}$ parameters by integration; and fourth remove the variances (plural because two more variances appear before we finish the calculation). Because the calculation is lengthy, we make many approximations of the kind that experienced users of applied mathematics learn to make. They could be avoided - but the calculation would then be much longer, with the same final conclusions.

We begin by the calculation in a manner similar to that done in Chapter 3. The question we would like to ask is "Given a set of model equations $\{f_1, \cdots, f_s\}$ and looking only within that set, which model best accounts for the data?" We will take

$$f_j(t) = \sum_{k=1}^{m} A_k H_k(t, \{\omega\})$$

as our model, where H_k are the orthonormal model functions defined earlier, Eq. (3.5). The subscripts "j" refers to the jth member of the set of models $\{f_1, \cdots, f_s\}$, with the understanding that the amplitudes $\{A\}$, the nonlinear $\{\omega\}$, the total number of model functions m, and the model functions $H_k(t, \{\omega\})$ are different for every f_j. We could label each of these with an additional subscript, for example H_{jk} to stand for model function k of model f_j; however, we will not do this simply because the proliferation of subscripts would render the mathematics unreadable.

To calculate the direct probability of the data given model f_j we take the difference between the data and the model. This difference is the noise, if the model is true, and making the most conservative assumptions possible about the noise we assign a Gaussian prior for the noise. This gives the

$$P(D|\{A\}, \{\omega\}, \sigma, f_j, I) = (2\pi\sigma^2)^{-\frac{N}{2}} \exp\left\{-\frac{1}{2\sigma^2}\sum_{i=1}^{N}[d_i - f_j(t_i)]^2\right\}$$

as the direct probability of the data given model f_j and the parameters. Now expanding the square we obtain

$$P(D|\{A\}, \{\omega\}, \sigma, f_j, I) = (2\pi\sigma^2)^{-\frac{N}{2}} \exp\left\{-\frac{NQ}{2\sigma^2}\right\}$$

where

$$Q \equiv \overline{d^2} - \frac{2}{N}\sum_{l=1}^{m} A_l h_l + \frac{1}{N}\sum_{l=1}^{m} A_l^2$$

and

$$\sum_{i=1}^{N} H_l(t_i) H_k(t_i) = \delta_{lk}$$

and

$$h_l = \sum_{i=1}^{N} d_i H_l(t_i)$$

was used to simplify the expression. This is now substituted back into Eq. (5.2) to obtain

$$P(D|\sigma, f_j, I) = \int d\{A\} d\{\omega\} P(\{A\}, \{\omega\}|I)(2\pi\sigma^2)^{-\frac{N}{2}} \exp\left\{-\frac{NQ}{2\sigma^2}\right\}. \tag{5.3}$$

At this point in the calculation we have simply repeated the steps done in Chapter 3 with one exception: we have retained the normalization constants in the direct probability. To remove the amplitudes we assign an appropriate normalized prior and integrate. When we compare models with the same number of amplitudes and the same priors for them, the prior normalization factors do not matter: they simply cancel out of the posterior probability (5.1). But when we compare a model to one that has fewer amplitudes, these prior factors no longer cancel. We must keep track of them. In the calculation in Chapter 3 we used an improper uniform prior for these parameters. We cannot do that here because it smears out our prior information over an infinite range, and this would automatically exclude the larger model.

We will assume that the parameters are logically independent in the sense that gaining information about the amplitudes $\{A\}$ will not change our information about the nonlinear $\{\omega\}$ parameters, thus the prior factors:

$$P(\{A\},\{\omega\}|I) = P(\{A\}|I)P(\{\omega\}|I). \tag{5.4}$$

The amplitudes are location parameters and in Appendix A we derived an appropriate informative prior for a location parameter: the Gaussian. We will assume we have a vague previous measurement of the amplitudes $\{A\}$ and express this as a Gaussian centered at zero. Thus we take

$$P(\{A\}|\delta,I) = (2\pi\delta^2)^{-\frac{m}{2}} \exp\left\{-\sum_{k=1}^{m} \frac{A_k^2}{2\delta^2}\right\} \tag{5.5}$$

as our informative prior. In the Bayesian literature, δ is called a "hyperparameter". We will do this calculation for the case where we have little (effectively no) prior information: we assume $\delta^2 \gg \sigma^2$. That is, the prior measurement is much worse than the current measurement. Then the orthonormal amplitudes $\{A\}$ are all estimated with the same precision δ as required by Eq. (4.4).

Substituting the factored prior, Eq. (5.4), into Eq. (5.3) and then substituting the prior, Eq. (5.5), into Eq. (5.3) we arrive at

$$\begin{aligned} P(D|\delta,\sigma,f_j,I) &= \int d\{\omega\}P(\{\omega\}|I)(2\pi\delta^2)^{-\frac{m}{2}}(2\pi\sigma^2)^{-\frac{N}{2}} \\ &\times \int_{-\infty}^{+\infty} dA_1 \cdots dA_m \exp\left\{-\sum_{k=1}^{m}\frac{A_k^2}{2\delta^2}\right\} \\ &\times \exp\left\{-\frac{1}{2\sigma^2}\left[N\overline{d^2} - 2\sum_{k=1}^{m} A_k h_k + \sum_{k=1}^{m} A_k^2\right]\right\} \end{aligned}$$

as the direct probability of the data given model function f_j and the parameters. What is essential here is that the prior may be considered a constant over the range of values where the likelihood is large, but it goes to zero outside that range fast enough to make it normalizable. Thus the last term in this integral looks like a delta function compared to the prior. We may write

$$\begin{aligned} P(D|\delta,\sigma,f_j,I) &= \int d\{\omega\}P(\{\omega\}|I)(2\pi\sigma^2)^{-\frac{N}{2}}(2\pi\delta^2)^{-\frac{m}{2}}\exp\left\{-\sum_{k=1}^{m}\frac{\hat{A}_k^2}{2\delta^2}\right\} \\ &\times \int_{-\infty}^{+\infty} d\{A\}\exp\left\{-\frac{1}{2\sigma^2}\left[N\overline{d^2}-2\sum_{k=1}^{m}A_k h_k+\sum_{k=1}^{m}A_k^2\right]\right\} \end{aligned}$$

where $\hat{A}$ is the location of the maximum of the likelihood as a function of the $\{A\}$ parameters. But from (4.1) $\hat{A}_j = h_j$ for a given model and after completing the integrals over the amplitudes, we have

$$\begin{aligned} P(D|\delta,\sigma,f_j,I) &= \int d\{\omega\}(2\pi\delta^2)^{-\frac{m}{2}}(2\pi\sigma^2)^{-\frac{(N-m)}{2}}P(\{\omega\}|I) \\ &\times \exp\left\{-\frac{N\overline{d^2}-m\overline{h^2}}{2\sigma^2}-\frac{m\overline{h^2}}{2\delta^2}\right\} \end{aligned}$$

as the direct probability of the data given the model function f_j and the parameters.

The second step in this calculation is to approximate $\overline{h^2}$ around the maximum and then perform the integrals over the $\{\omega\}$ parameters. The prior uncertainty $\delta \gg \sigma$, so the prior factor in the above equation is only a small correction. When we expand $\overline{h^2}$ about the maximum $\{\hat{\omega}\}$ we will not bother expanding this term. This permits us to use the same approximation given earlier (4.10, 4.9) while making only a small error. We Taylor expand $\overline{h^2}$ to obtain

$$\begin{aligned} P(D|\delta,\sigma,f_j,I) &\approx \int d\{\omega\}P(\{\omega\}|I)(2\pi\delta^2)^{-\frac{m}{2}}(2\pi\sigma^2)^{-\frac{N-m}{2}} \\ &\times \exp\left\{-\frac{N\overline{d^2}-m\overline{h^2}(\{\hat{\omega}\})}{2\sigma^2}-\frac{m\overline{h^2}(\{\hat{\omega}\})}{2\delta^2}\right\} \\ &\times \exp\left\{-\sum_{k,l}^{r}\frac{b_{kl}(\hat{\omega}_k-\omega_k)(\hat{\omega}_l-\omega_l)}{2\sigma^2}\right\}. \end{aligned} \tag{5.6}$$

We are now in a position to remove the $\{\omega\}$ parameters. To do this third step in the calculation we will again assign a normalized prior for them. When we Taylor expanded $\overline{h^2}$ we made a local approximation to the direct probability of the data given the parameters. In this approximation the $\{\omega\}$ parameters are location parameters. We again assume a prior which is Gaussian with some variance γ, another hyperparameter. We have

$$P(\{\omega\}|\gamma,I) = (2\pi\gamma^2)^{-\frac{r}{2}}\exp\left\{-\sum_{k=1}^{r}\frac{\omega_k^2}{2\gamma^2}\right\} \tag{5.7}$$

as the informative prior for the $\{\omega\}$ parameters. If the $\{\omega\}$ parameters are frequencies then one could argue that they are scale parameters, for which the completely uninformative prior is the nonnormalizable Jeffreys prior; and so we should choose a normalizable prior that resembles it. However, that does not matter; the only properties of our prior that survive are the prior density at the maximum likelihood point and the prior range, and even these may cancel out in the end. We are simply playing it safe by using normalized priors so that no singular mathematics can arise in our calculation; and it does not matter which particular ones we use as long as they are broad and uninformative.

Substituting the prior (5.7) into Eq. (5.6), the integral we must perform becomes

$$\begin{aligned} P(D|\gamma,\delta,\sigma,f_j,I) \approx & \int d\{\omega\}(2\pi\delta^2)^{-\frac{m}{2}}(2\pi\gamma^2)^{-\frac{r}{2}}(2\pi\sigma^2)^{-\frac{N-m}{2}} \\ & \times \exp\left\{-\frac{N\overline{d^2}-m\overline{h^2}(\{\hat{\omega}\})}{2\sigma^2}-\frac{m\overline{h^2}(\{\hat{\omega}\})}{2\delta^2}-\sum_{k=1}^{r}\frac{\omega_k^2}{2\gamma^2}\right\} \\ & \times \exp\left\{-\sum_{k,l}^{r}\frac{b_{kl}(\hat{\omega}_k-\omega_k)(\hat{\omega}_l-\omega_l)}{2\sigma^2}\right\}. \end{aligned}$$

We will again assume that the prior information is vague, $\gamma \gg \sigma$, we treat the last term in the integral like a delta function compared to the prior. Thus we will take the prior factors out of the integral and simply evaluate them at the maximum likelihood point. Then integrating over the $\{\omega\}$ parameters gives the direct probability of the data given the model f_j and the three remaining parameters. If these three parameters are actually known then the direct probability is given by

$$\begin{aligned} P(D|\gamma,\delta,\sigma,f_j,I) = & (2\pi\delta^2)^{-\frac{m}{2}}\exp\left\{-\frac{m\overline{h^2}(\{\hat{\omega}\})}{2\delta^2}\right\} \\ & \times (2\pi\gamma^2)^{-\frac{r}{2}}\exp\left\{-\frac{r\overline{\omega^2}}{2\gamma^2}\right\}v_1^{-\frac{1}{2}}\cdots v_r^{-\frac{1}{2}} \\ & \times (2\pi\sigma^2)^{-\frac{N-m-r}{2}}\exp\left\{-\frac{N\overline{d^2}-m\overline{h^2}(\{\hat{\omega}\})}{2\sigma^2}\right\} \end{aligned} \quad (5.8)$$

where $\overline{\omega^2} = (1/r)\sum_{k=1}^{r}\hat{\omega}^2$ is the mean-square $\{\hat{\omega}\}$ for model f_j, and $\overline{h^2}(\{\hat{\omega}\})$ is the mean-square projection of the data onto the orthonormal model functions evaluated at the maximum likelihood point for model f_j, and $v_1^{-\frac{1}{2}}\cdots v_r^{-\frac{1}{2}}$ is the Jacobian introduced in Eq. (4.12). If the three variances are known then the problem is complete, and the number which must be used in (5.1) is given by (5.8).

We noted earlier that one must be careful with the prior factors when the models have different numbers of parameters and we can see that here. If two models have different values of m or r, their relative likelihood will have factors of the form $(2\pi\delta^2)^x$, or $(2\pi\gamma^2)^y$. The prior ranges remain relevant, a fact that we would have missed had we used improper priors.

There are three variances, σ, δ, and γ, in the direct probability of the data. We would like to remove these from $P(D|\gamma, \delta, \sigma, f_j, I)$. We could remove these using a Jeffreys prior, because each of these parameters appears in every model. The infinity introduced in doing this would always cancel out formally. However, to be safe, we can bound the integral, normalize the Jeffreys prior, and then remove these variances; then even if the normalization constant did not cancel we would still obtain the correct result. We will proceed with this last approach. There are three variances, and therefore three integrals to perform. Each of the three integrals is of the form:

$$\frac{1}{\log(H/L)} \int_L^H ds \frac{s^{-a} \exp\left\{-\frac{Q}{s^2}\right\}}{s}$$

where H stands for the upper bound on the variance, L for the lower bound, $\log(H/L)$ is the normalization constant for the Jeffreys prior, s is any one of the three variances, and Q and a are constants associated with s. A change of variables $u = Q/s^2$ reduces this integral to

$$\frac{1}{2} \frac{Q^{-\frac{a}{2}}}{\log(H/L)} \int_{\sqrt{\frac{Q}{H}}}^{\sqrt{\frac{Q}{L}}} du\, u^{\frac{a}{2}-1} e^{-u}.$$

This integral is of the form of an incomplete Gamma integral. But our knowledge of the limits on this integral is vague: we know only that L is small and H large. If, for example, we assume that

$$\sqrt{\frac{Q}{H}} \ll 1 \quad \text{and} \quad \frac{a}{2} - 1 \ll \sqrt{\frac{Q}{L}}$$

then the integrand is effectively zero at the limits; we can take the integral to be approximately $\Gamma(a/2)$. Designating the ratio of the limits H/L as R_α, where the subscript represents the limits for σ, δ, or γ integral, the three integrals are given approximately by

$$\int_L^H d\delta \frac{\delta^{-m} \exp\left\{-m\overline{h^2}/2\delta^2\right\}}{\log(R_\delta)\delta} \approx \frac{\Gamma(m/2)}{2\log(R_\alpha)} \left[\frac{m\overline{h^2}}{2}\right]^{-\frac{m}{2}}$$

for δ and

$$\int_L^H d\gamma \frac{\gamma^{-r} \exp\left\{-r\overline{\omega^2}/2\gamma^2\right\}}{\log(R_\gamma)\gamma} \approx \frac{\Gamma(r/2)}{2\log(R_\gamma)} \left[\frac{r\overline{\omega^2}}{2}\right]^{-\frac{r}{2}}$$

for γ and

$$\int_L^H d\sigma \frac{\sigma^{m+r-N} \exp\left\{-[N\overline{d^2} - m\overline{h^2}]/2\sigma^2\right\}}{\log(R_\sigma)\sigma} \approx \frac{\Gamma(\frac{N-m-r}{2})}{2\log(R_\sigma)} \left[\frac{N\overline{d^2} - m\overline{h^2}(\{\omega\})}{2}\right]^{\frac{m+r-N}{2}}$$

for σ.

Using these three integrals the global likelihood of the data given the model f_j is given by

$$\begin{aligned} P(D|f_j, I) &= \frac{\Gamma(m/2)}{2\log(R_\delta)} \left[\frac{m\overline{h^2}(\{\hat{\omega}\})}{2}\right]^{-\frac{m}{2}} \frac{\Gamma(r/2)}{2\log(R_\gamma)} \left[\frac{r\overline{\omega^2}}{2}\right]^{-\frac{r}{2}} v_1^{-\frac{1}{2}} \cdots v_r^{-\frac{1}{2}} \\ &\times \frac{\Gamma([N-m-r]/2)}{2\log(R_\sigma)} \left[\frac{N\overline{d^2} - m\overline{h^2}(\{\omega\})}{2}\right]^{\frac{m+r-N}{2}}. \end{aligned} \tag{5.9}$$

The three factors involved in normalizing the Jeffreys priors appear in every model: they always cancel as long as we deal with models having all three types of parameters. However, as soon as we try to compare a model involving two types of parameters to a model involving all three types of parameters (e.g. a regression model to a nonlinear model) they no longer cancel: the prior ranges become important. One must think carefully about just what prior information one actually has about γ, and δ, and use that information to set their prior ranges. As we shall see in what follows, if the data actually determine the model parameters well (so that these equations apply) the actual values one assigns to δ and γ are relatively unimportant.

5.3 One More Parameter

We would like to understand (5.1), (5.8), and (5.9) better, and we present here a simple example of their use. Suppose we are dealing with the simplest model selection possible: expanding the data on a set of model functions. A typical set of model functions might be polynomials. This is the simplest model possible because there are no $\{\omega\}$ parameters; only amplitudes. But suppose further that we choose our model functions so that they are already orthogonal in the sense defined earlier. All that is left for us to decide is "When have we incorporated enough expansion vectors to adequately represent the signal?" We will assume in this demonstration that both the variance σ and the prior variance δ are known and apply (5.8). Further, we will be comparing only two models at a time, and will compute the ratio of Eq. (5.8) for a model containing m expansion functions (or vectors on the discrete sample points) to a model containing $m+1$ expansion functions. This ratio is called the posterior "odds" in favor of the smaller model.

When we have m expansion vectors and no $\{\omega\}$, Eq. (5.8) reduces to

$$\begin{aligned} P(D|f_m,\sigma,\delta,I) &= (2\pi\delta^2)^{-\frac{m}{2}}\exp\left\{-\sum_{k=1}^{m}\frac{h_k^2}{2\delta^2}\right\} \\ &\times (2\pi\sigma^2)^{-\frac{N-m}{2}}\exp\left\{-\frac{1}{2\sigma^2}\left[N\overline{d^2}-\sum_{k=1}^{m}h_k^2\right]\right\} \end{aligned}$$

for the first model and to

$$\begin{aligned} P(D|f_{m+1},I) &= (2\pi\delta^2)^{-\frac{m+1}{2}}\exp\left\{-\sum_{k=1}^{m+1}\frac{h_k^2}{2\delta^2}\right\} \\ &\times (2\pi\sigma^2)^{-\frac{N-m-1}{2}}\exp\left\{-\frac{1}{2\sigma^2}\left[N\overline{d^2}-\sum_{k=1}^{m+1}h_k^2\right]\right\} \end{aligned}$$

for a model with $m+1$ parameters. Because these models are already orthonormal, h_k is the same in both equations: when we compute the odds ratio all but the last will cancel. Thus the posterior odds ratio simplifies considerably. We have the likelihood ratio

$$L=\frac{P(D|f_m,\sigma,\delta,I)}{P(D|f_{m+1},\sigma,\delta,I)}=\frac{\delta}{\sigma}\exp\left\{\left(\frac{1}{\delta^2}-\frac{1}{\sigma^2}\right)\frac{h_{m+1}^2}{2}\right\}.$$

The posterior odds ratio then involves the posterior probabilities:

$$\frac{P(f_m|\sigma,\delta,D,I)}{P(f_{m+1}|\sigma,\delta,D,I)}=\frac{P(f_m|I)}{P(f_{m+1}|I)}L.$$

We derived this approximation assuming $\delta \gg \sigma$, so we have

$$L=\frac{\delta}{\sigma}\exp\left\{-\frac{h_{m+1}^2}{2\sigma^2}\right\}.$$

In other words, the smaller model is helped by uncertainty in the prior knowledge of A_{m+1}, while the larger model is helped by the relative size of the estimated next amplitude compared to the noise. This is the Bayesian quantitative version of Occam's razor: prefer the simpler model unless the bigger one achieves a significantly better fit to the data. For the bigger model to be preferred, the $m+1$ model function's projection onto the data must be large compared to the noise. Thus the Bayesian answer to this question essentially tells one to do what his common sense might have told him to do. That is, to continue increasing the number of expansion vectors until the projection of the data onto the next vector becomes comparable to the noise.

But we can be more specific than this. For example assume that $100\sigma = \delta$. Then to achieve $L = 1$, we need

$$\begin{aligned} \log(100) - \frac{h_{m+1}^2}{2\sigma^2} &= 0 \\ h_{m+1} &= \pm 3.0\sigma. \end{aligned}$$

The "data fitting factor" cancels out the "Occam factor" when the next projection is three times the RMS noise. Projections larger than this will favor the more complicated model.

This result does not depend strongly on the assumed prior information. Here we took δ to be 100 times larger than σ. But the answer depends on the square root of $\log(\delta)$. So even if δ had been a billion (10^9) times larger it would have increased the critical value of h_{m+1} only by a factor of 2.3. Thus probability theory can give one a reasonable criterion for choosing between models, that depends only weakly on the prior information. There is hardly any real problem in which one would not feel sure in advance that $\delta < 10^{11}\sigma$, and few in which that 10^{11} could not be reduced to 10^2. But to try to go an improper prior $\delta \to \infty$, would give entirely misleading results; the larger model could never be accepted, whatever the data. Thus, while use of proper priors is irrelevant in many problems, it is mandatory in some.

5.4 What is a Good Model?

We can now state what we mean by a good model. We know from the Bessel inequality (3.17) that the estimated noise variance will have a value of $\overline{d^2}$ when we have no model functions. As we include more model functions, the estimated variance must go monotonically to zero. We can plot the estimated variance as a function of the expansion order, Fig. 5.1 (by expansion order we mean the total number of model functions m).

There are three general regions of interest: First, the solid line running from $\overline{d^2}$ down to zero (we will call this a "bad" model); second the region with values of σ^2 below this line; and third the region above this line. The region above the line is not a bad or a good region; it is simply one in which the model functions have been labeled in a bad order. By reordering the model functions we will obtain a curve below the straight line.

Figure 5.1: Choosing a Model

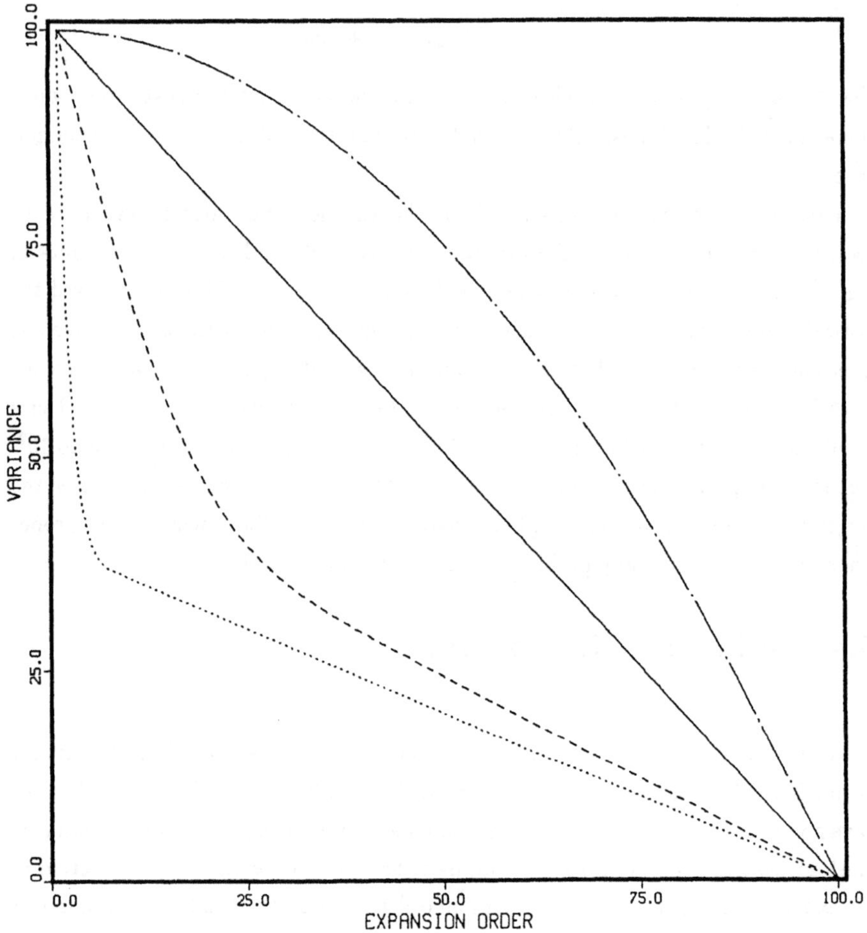

The solid line represents the worst possible choice of model functions. The region above this line is neither good nor bad (see text). The region below the line represents the behavior of good models. One strives to obtain the largest drop in the estimated variance with the fewest model functions. The dashed line might represent a fair model and the dotted line the "best" model.

Let $\Delta(\langle\sigma^2\rangle)$ stand for the change in the estimated variance σ^2 from incorporating one additional model function [we define $\Delta(\langle\sigma^2\rangle)$ to be positive]. We assume the model functions are incorporated in order of decreasing $\Delta(\langle\sigma^2\rangle)$: the model function with the largest $\Delta(\langle\sigma^2\rangle)$ is labeled one; the model function which produces the second largest $\Delta(\langle\sigma^2\rangle)$ is number two, etc.

We called the solid line a "bad" model because all of the $\Delta(\langle\sigma^2\rangle)$'s are the same; there is no particular model function which resembles the data better than any other. It would require outstandingly bad judgment – or bad luck – to choose such a set of model functions. But something like the linear behavior is to be expected when one expands pure noise on a complete set. On the other hand, if there is a signal present one expects to do better than this until the signal has been expanded; then one expects the curve to become slowly varying.

We can characterize a model by how quickly the $\Delta(\langle\sigma^2\rangle)$ drops. Any curve which drops below another curve indicates a model which is better, in the sense that it achieves a better quality of fit to the data with a given number of model functions. The "best" model is one which projects as much mean-square data as possible onto the first few model functions. What one would expect to find is: a very rapid drop as the systematic signal is taken up by the model, followed by a slow drop as additional model functions expand the noise.

We now have the following intuitive picture of the model fitting process: one strives to find models which produce the largest and fastest drop in $\langle\sigma^2\rangle$; any model which absorbs the systematic part of the signal faster than another model is a better model; the "best" is one which absorbs all of the systematic part of the signal with the fewest model parameters. This corresponds to the usual course of a scientific research project; initially one is very unsure of the phenomenon and so allows many conceivable unknown parameters with a complicated model. With experience one learns which parameters are irrelevant and removes them, giving a simpler model that accounts for the facts with fewer model functions. The total number of "useful" model functions is determined by the location of the break in the curve. The probability of any particular model can be computed using (4.15), and this can be used to estimate where the break in the curve occurs.

Of course, in a very complicated problem, where the data are contaminated by many spurious features that one could hardly hope to capture in a model, there may not be any well-defined breaking point. Even so, the curve is useful in that its shape gives some indication of how clean-cut the problem is.

Chapter 6

SPECTRAL ESTIMATION

The previous chapters surveyed the general theory. In this chapter we will specialize the analysis to frequency and spectrum estimates. Our ultimate aim is to derive explicit Bayesian estimates of the power spectrum and other parameters when multiple nonstationary frequencies are present. We will do this by proceeding through several stages beginning with the simplest spectrum estimation problem. We do this because as was shown by Jaynes [12] when multiple well-separated frequencies are present $[|\omega_j - \omega_k| \gg 2\pi/N]$, the spectrum estimation problem essentially separates into independent single-frequency problems. It is only when multiple frequencies are close together that we will need to use more general models.

In Chapters 3 and 4 we derived the posterior probability of the $\{\omega\}$ parameters independent of the amplitudes and noise variance and without assuming the sampling times t_i to be uniformly spaced. Much of the discussion in this Chapter will center around understanding the behavior of the posterior probability density for multiple frequencies. This discussion is, of course, simpler when the t_i are uniform, because then the sine and cosine terms are orthogonal in the sense discussed before. We will start by making this assumption; then, where appropriate, the results for nonuniform times will be given.

This should not be taken to imply that uniform time spacing is the "best" way to obtain data. In fact, nonuniform time intervals have some significant advantages over uniform intervals. We will discuss this issue shortly, and show that obtaining data at apparently random intervals will significantly improve the discrimination of high frequencies even with the same amount of data.

6.1 The Spectrum of a Single Frequency

In Chapter 2 we worked out an approximate Bayesian solution to the single stationary harmonic frequency problem. Then in Chapter 3 we worked out what amounts to the general solution to this problem. Because we have addressed this problem so thoroughly in other places we will investigate some other properties of the analysis that may be troubling the reader. In particular we would like to understand what happens to the ability to estimate parameters when one or more of our assumptions is violated. We would like to demonstrate that the estimates derived in Chapter 4 are accurate, and that when the assumptions are violated the estimated frequencies are still reasonably correct but the error estimates are larger, and therefore, more conservative.

6.1.1 The "Student t-Distribution"

We begin this chapter by demonstrating how to use the general formalism to derive the exact "Student t-distribution" for the single frequency problem on a uniform grid. For a uniformly sampled time series, the model equation is

$$f_l = B_1 \cos \omega l + B_2 \sin \omega l$$

where l is an index running over a symmetric time interval $(-T \leq l \leq T)$ and $(2T + 1 = N)$. The matrix g_{ij}, Eq. (3.4), becomes

$$g_{ij} = \begin{pmatrix} \sum_{l=-T}^{T} \cos^2 \omega l & \sum_{l=-T}^{T} \cos \omega l \sin \omega l \\ \sum_{l=-T}^{T} \cos \omega l \sin \omega l & \sum_{l=-T}^{T} \sin^2 \omega l \end{pmatrix}.$$

For uniform time sampling the off diagonal terms are zero and the diagonal term may be summed explicitly to obtain

$$g_{ij} = \begin{pmatrix} c & 0 \\ 0 & s \end{pmatrix}$$

where

$$c = \frac{N}{2} + \frac{\sin(N\omega)}{2\sin(\omega)}$$

$$s = \frac{N}{2} - \frac{\sin(N\omega)}{2\sin(\omega)}.$$

Then the orthonormal model functions may be written as

$$H_1(t) = \frac{\cos(\omega t)}{\sqrt{c}}$$

$$H_2(t) = \frac{\sin(\omega t)}{\sqrt{s}}.$$

The posterior probability of a frequency ω in a uniformly sampled data set is given by Eq. (3.17). Substituting these model functions gives

$$P(\omega|D,I) \propto \left[1 - \frac{R(\omega)^2/c + I(\omega)^2/s}{N\overline{d^2}}\right]^{\frac{2-N}{2}} \tag{6.1}$$

where $R(\omega)$ and $I(\omega)$ are the squares of the real and imaginary parts of the discrete Fourier transform (2.4, 2.5).

We see now why the discrete Fourier transform does poorly for small N or low frequencies: the constants c and s are normalization constants that usually reduce to $N/2$ for large N; however, these constants can vary significantly from $N/2$ for small N or low frequency. Thus the discrete Fourier transform is only an approximate result that must be replaced by (6.1) for small amounts of data or data sets which contain a low frequency. The general solution is represented by (3.17), and this equation may be applied even when the sampling is nonuniform.

6.1.2 Example – Single Harmonic Frequency

To obtain a better understanding of the use of the power spectral density derived in Chapter 2 (2.16), we have prepared an example: the data consist of a single harmonic frequency plus Gaussian white noise, Fig. 6.1. We generated these data from the following equation

$$d_j = 0.001 + \cos(0.3j + 1) + e_j \tag{6.2}$$

where j is a simple index running over the symmetric interval $-T$ to T in integer steps ($2T+1 = 512$), and e_j is a random number with unit variance. After generating the time series we computed its average value and subtracted it from each data point: this ensures that the data have zero mean value. Figure 6.1(A) is a plot of this computer simulated time series, and Fig. 6.1(B) is a plot of the Schuster periodogram (continuous curve) with the fast Fourier transform marked with open circles. The

Figure 6.1: Single Frequency Estimation

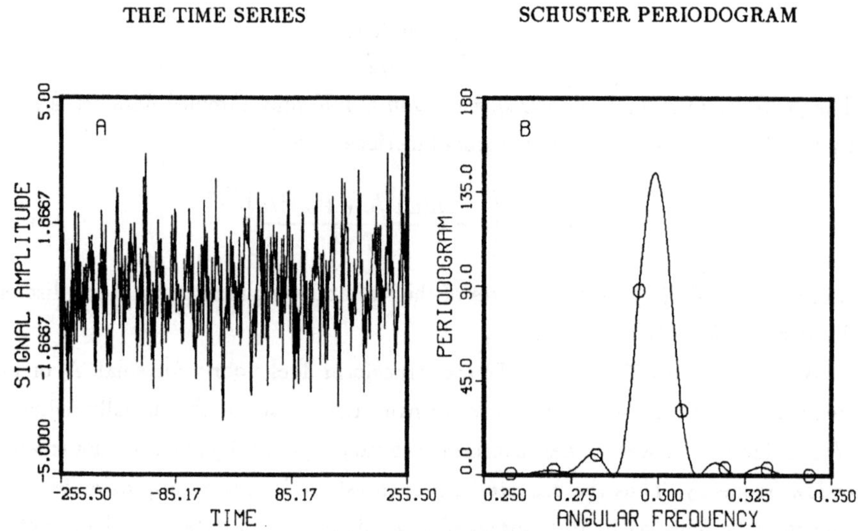

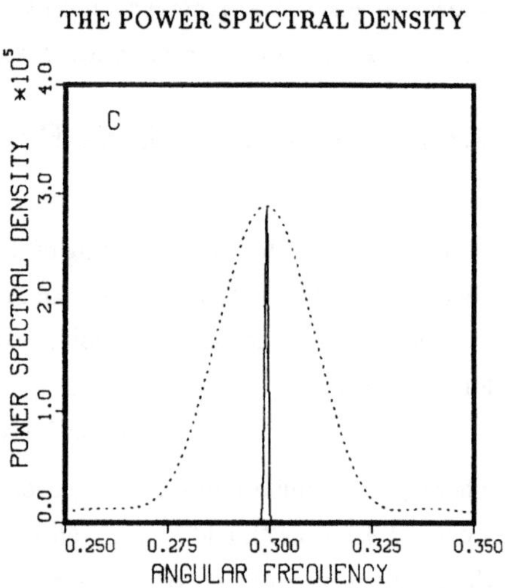

The data (A) contain a single harmonic frequency plus noise; $N = 512$, and $S/N \approx 1$. The Schuster periodogram, (B) solid curve, and the fast Fourier transform, open circles, clearly show a sharp peak plus side lobes. These side lobes do not show up in the power spectral density, (C), because the posterior probability is very sharply peaked around the maximum of the periodogram. The dotted line in (C) is a Blackman-Tukey spectrum with a Hanning window and 256 lag coefficients.

periodogram and the fast Fourier transform have spurious side lobes, but these do not appear in the plot of the power spectral density estimate, Fig. 6.1(C), because as noted in Chapter 2, the processing in (4.15) will effectively suppress all but the very highest peak in the periodogram. This just illustrates numerically what we already knew analytically; it is only the very highest part of the periodogram that is important for estimation of a single frequency.

We have included a Blackman-Tukey spectrum using a Hanning window (dotted line) in Figure 6.1(C) for comparison. The Blackman-Tukey spectrum has removed the side lobes at the cost of half the resolution. The maximum lag was set at 256, i.e. over half the data. Had we used a lag of one-tenth as Tukey [13] advocates, the Blackman-Tukey spectrum would look nearly like a horizontal straight line on the scale of this plot.

Of course, the peak of the periodogram and the peak of the power spectral density occur at the same frequency. Indeed, for a simple harmonic signal the peak of the periodogram is the optimum frequency estimator. But in our problem (i.e. our model), the periodogram is not even approximately a valid estimator of the power spectrum, as we noted earlier. Consequently, even though these techniques give nearly the same frequency estimates, they give very different power spectral estimates and, from the discussion in Chapters 2 and 4, very different accuracy estimates.

Probably, one should explain the difference on the grounds that the two procedures are solving different problems. Unfortunately, we are unable to show this explicitly. We have shown above in detail that the Bayesian procedure yields the optimal solution to a well-formulated problem, by a well-defined criterion of optimality. One who wishes to solve a different problem, or to use a different optimality criterion, will naturally seek a different procedure. The Blackman-Tukey procedure has not, to the best of our knowledge, been so related to any specific problem, much less to any optimality criterion; it was introduced as an intuitive, *ad hoc* device. We know that Blackman and Tukey had in mind the case where the entire time series is considered a sample drawn from a "stationary Gaussian random process"; thus it has no mention of such notions as "signal" and "noise". But the "hanning window" smoothing procedure has no theoretical relation to that problem; and of course the Bayesian solution to it (given implicitly by Geisser and Cornfield [24] and Zellner [25] in their Bayesian estimates of the covariance matrix of a multivariate Gaussian sampling distribution) would be very different. It is still conceivable that the Blackman-Tukey procedure is the solution to some well-defined problem, but we do not know what that problem is.

6.1.3 The Sampling Distribution of the Estimates

We mentioned in Chapter 4 that we would illustrate numerically that the Bayesian estimates for the $\{\omega\}$ parameters were indeed accurate under the conditions supposed, even by sampling-theory criteria. For the example just given the true frequency was 0.3 while the estimated frequency from data with unity signal-to-noise ratio was

$$(\omega)_{\text{est}} = 0.2997 \pm 0.0006$$

at two standard deviations, in dimensionless units. But one example in which the estimate is accurate is not a sufficient demonstration. Suppose we generate the signal (6.2) a number of times and allow the noise to be different in each of these. We can then compute a histogram of the number of times the frequency estimate was within one standard deviation, two standard deviations $\cdots$ etc. of the true value. We could then plot the histogram and compare this to a Gaussian, or we could integrate the histogram and compare the total percentage of estimates included in the interval $\hat{\omega} - \langle\omega\rangle$ to a Gaussian. This would tell us something about how accurately the results are reproducible over different noise samples. This is not the same thing as the accuracy with which ω is estimated from one given data set; but orthodox statistical theory takes no note of the distinction. Indeed, in "orthodox" statistical theory, this sampling distribution of the estimates is the sole criterion used in judging the merits of an estimation procedure.

We did this numerically by generating some 3000 samples of (6.2) and estimating the frequency ω from each one. We then computed the histogram of the estimates, integrated, and plotted the total percentage of estimates enclosed as a function of $\hat{\omega} - \langle\omega\rangle$, Fig. 6.2 (solid line). From the 3000 sample estimates we computed the mean and standard deviation. The dashed line is the equivalent plot for a Gaussian having this mean and standard deviation. With 3000 samples the empirical sampling distribution is effectively identical to this Gaussian, and its width corresponds closely to the Bayesian error estimate. However, as R. A. Fisher explained many years ago, this agreement need not hold when the estimator is not a sufficient statistic.

6.1.4 Violating the Assumptions – Robustness

We have said a number of times that the estimates we are making are the "most conservative" estimates of the parameters one can make. We would like to convey a

Figure 6.2: The Distribution of the Sample Estimates

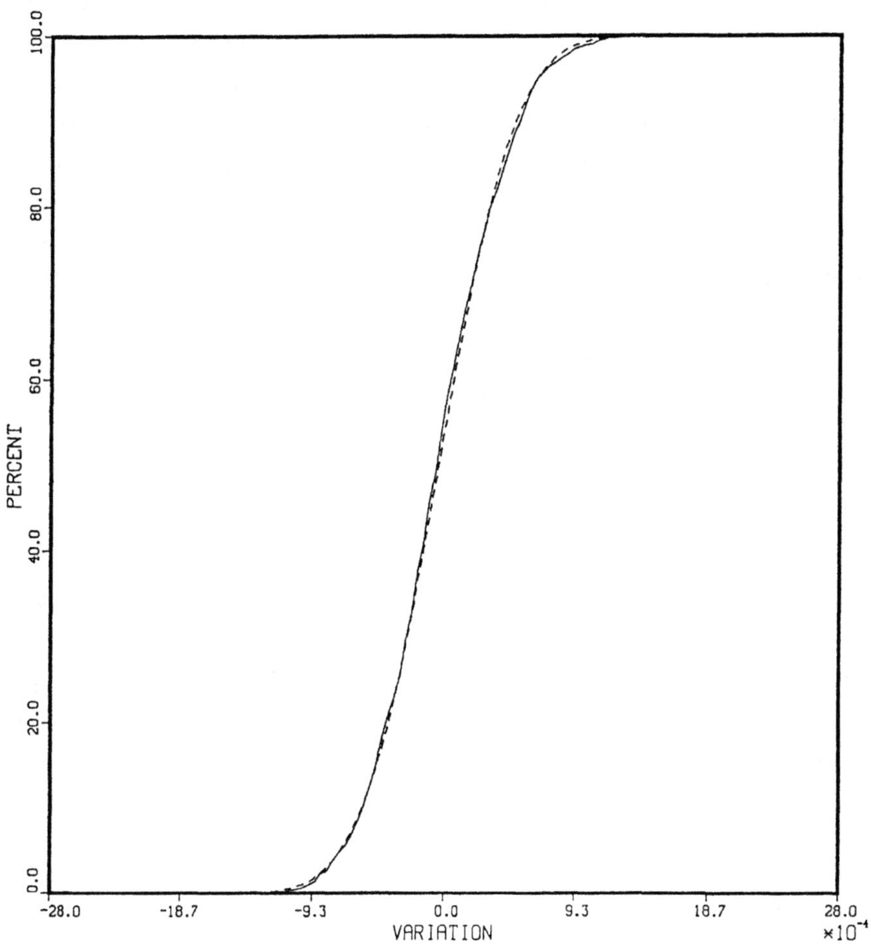

We generated the single harmonic signal (6.2) some 3000 times and estimated the frequency. We computed the mean, standard deviation, and a histogram from these data, then totaled the number of estimates from left to right; this gives the total percentage of estimates enclosed as a function of $\hat{\omega} - \langle \omega \rangle$ (solid line). We have plotted an equivalent Gaussian (dashed line) having the same mean and standard deviation as the sample. Each tick mark on the plot corresponds to two standard deviations.

better understanding of that term now. General theorems guarantee [25], [26], [27], [28] that if all of the assumptions are met exactly, then the estimate we obtain will be the "best" estimate of the parameters that one can make from the data and the prior information. But in all cases where we had to put in prior information, we specifically assumed the least amount of information possible. This occurred when we chose a prior for the noise – we used maximum entropy to derive the most uninformative prior we could for a given second moment: the Gaussian. It occurred again when we assigned the priors for the amplitudes, and again when we assigned the prior for the $\{\omega\}$ parameters. This means that any estimate that takes into account additional information by using a more concentrated prior will always do better! But, further if the model assumptions are not met by the data (e.g. the noise is not white, the "true" signal is different from our model, etc.), then probability theory will necessarily make the accuracy estimates even wider because the models do not fit the data as well! These are bold claims, and we will demonstrate them for the single frequency model.

Periodic but Nonharmonic Signals

First let us investigate what will happen if the true signal in the data is different from that used in the model (i.e. it does not belong to the class of model functions assumed by the model). Consider the time series given in Fig. 6.3(A); this signal is a series of ramp functions. We generated the data with $N = 1024$ data points by simply running a counter from zero to 15, and repeated this process 64 times. The RMS is then $[1/16\sum_{k=0}^{15}(k-7.5)^2]^{\frac{1}{2}} = 4.61$. We then added a unit normal random number to the data, and last we computed the average value of the data and subtracted this from each data point.

This signal is periodic but not harmonic; nonetheless we propose to use the single harmonic frequency model on these data. Figure 6.3(B) is a plot of the $\log_{10}$ of the probability of a single harmonic frequency in these data: essentially this is the discrete Fourier transform of the data. We see in Fig. 6.3(B) the discrete Fourier transform has at least four peaks. But we have demonstrated that the discrete Fourier transform is an optimal frequency estimator for a single harmonic frequency: all of the structure, except the main peak, is a spurious artifact of not using the true model. The main peak in Fig. 6.3(B) is some 25 orders of magnitude above the second largest peak: probability theory is telling us all of that other structure is completely negligible. We then located this frequency as accurately as possible and computed the estimated

Figure 6.3: Periodic but Nonharmonic Time Signals

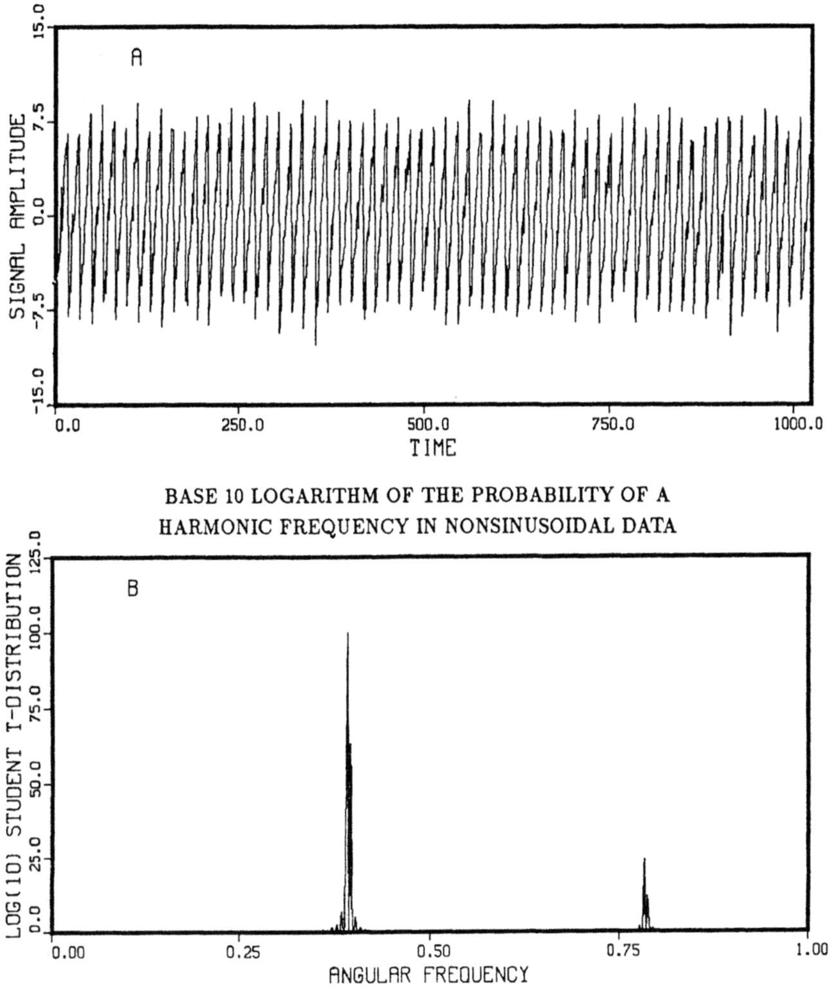

The data in (A) contain a periodic but nonharmonic frequency, with $N = 1024$, and $S/N \approx 4.6$. The Schuster periodogram, (B), clearly indicates a single sharp peak plus a number of other spurious features. Estimating the frequency from the peak of the periodogram gives 0.3927 ± 0.0003 while the true frequency is 0.3927.

error in the frequency using (4.14). This gives

$$(\omega)_{\text{est}} = 0.3927 \pm 0.0003$$

at two standard deviations. The true frequency is

$$(\omega)_{\text{true}} = 0.392699$$

while the "best" estimate possible for a sinusoidal signal with the same total number of data values and the same signal-to-noise would be given by

$$(\omega)_{\text{best}} = 0.392699 \pm 0.00003.$$

The estimate is a factor of ten worse than what could be obtained if the true signal met the assumptions of the model (i.e. was sinusoidal with the same signal-to-noise ratio). The major difference is in the estimated noise: the true signal-to-noise is 4.6, but the estimated signal-to-noise using the harmonic model is only 1.5.

The Effect of Nonstationary, Nonwhite Noise

What will be the effect of nonwhite noise on the ability to estimate a frequency? In preparing this test we used the same harmonic signal as in the simple harmonic frequency case (6.2). Although the noise is still Gaussian, we made it different from independent, identically distributed (iid) Gaussian in two ways: first, we made the noise increase linearly in time and second, we filtered the noise with a 1-2-1 filter. Thus the noise values not only increased in time; they were also correlated. The data for this example are shown in Fig. 6.4(A), and the $\log_{10}$ of the "Student t-distribution" is shown in Fig. 6.4(B). The data were prepared by first generating the simple harmonic frequency. We then prepared the noise using a Gaussian distributed random number generator, scaling linearly with increasing time, and filtering; finally, we added the noise to the data. The noise variance in these data ranges from 0.1 in the first data values to 2.1 in the last few data values – there are $N = 1000$ data values. We next computed the $\log_{10}$ probability of a single harmonic frequency in the data set, Fig. 6.4(B). There are two close peaks near 0.3 in dimensionless units. However, we now know that only the highest peak is important for frequency estimation. The highest peak is some 10 orders of magnitude above the second. Thus the second peak is completely negligible compared to the first. We estimated the frequency from this peak and found 0.297 ± 0.003; the correct value is 0.3. Thus one pays a penalty in

Figure 6.4: The Effect of Nonstationary, Nonwhite Noise

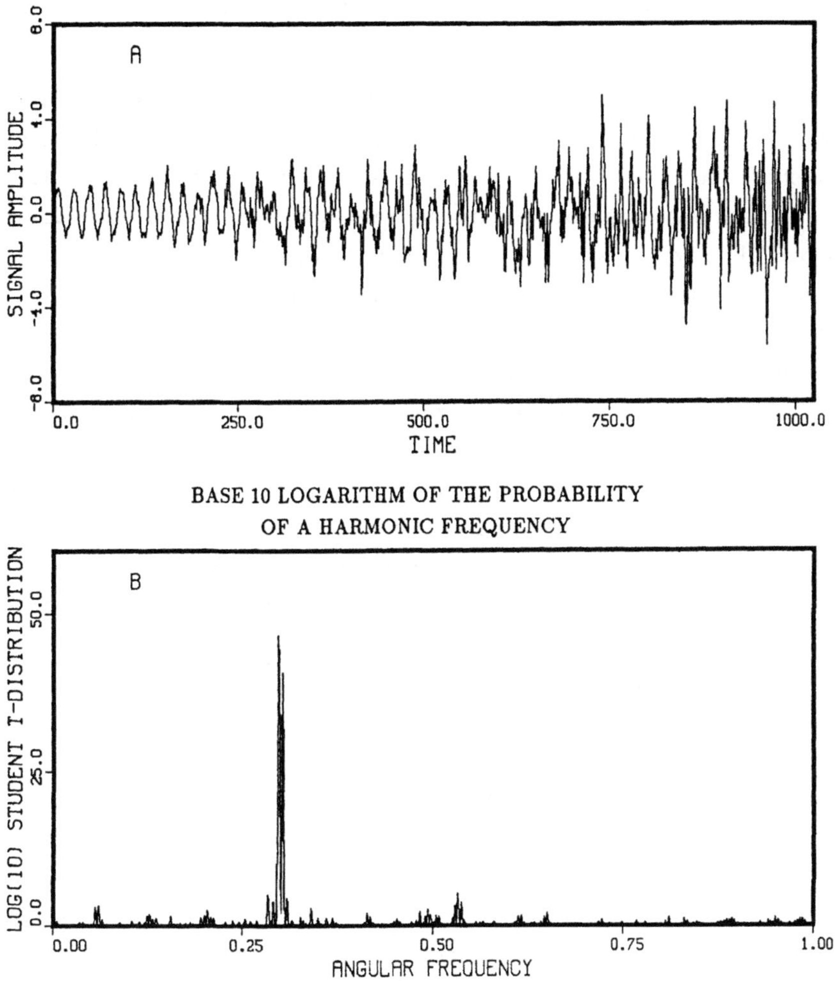

The data in Fig. 6.4(A) contain a periodic frequency but the noise is nonstationary and nonwhite, as described in the text. There are 1000 data points with $S/N < 0.5$, The Schuster periodogram, Fig. 6.4(B), clearly indicates a single sharp peak from which we estimated the frequency to be 0.297 ± 0.003; the correct value is 0.3.

accuracy; but the Bayesian conclusions still do not mislead us about this. Actually, the nonstationarity, which obscures part of the data, was much more serious than the nonwhiteness.

Amplitude modulation and other violations of the assumptions

It should be relatively clear by now what will happen when the amplitude of the signal is not constant. For the single stationary frequency problem the sufficient statistic is the Schuster periodogram, and we know from past experience that this statistic is at least usable on nonstationary series with Lorentzian or Gaussian decay. We can also say that when the amplitude modulation is completely unknown, the single largest peak in the discrete Fourier transform is the only indication of frequencies: all others are evidence but not proof. If one wishes to investigate these others one must include some information about the amplitude modulation.

It should be equally obvious that when the signal consists of several stationary sinusoids, the periodogram continues to work well as long as the frequencies are reasonably well separated. But any part of the data that does not fit the model is noise. In cases where we analyze data that contain multiple stationary frequencies using a one-frequency model, all of the frequencies except the one corresponding to the largest peak in the discrete Fourier transform are from the standpoint of probability theory just noise – and extremely correlated, non-Gaussian noise.

All of these effects, and why probability theory continues to work after a fashion in spite of them, are easily understood in terms of the intuitive picture given earlier on page 36. We are picking the frequency so that the dot product between the data and the model is as large as possible. In the case of the sawtooth function described earlier, it is obvious that the "best" fit will occur when the frequency matches that of the sawtooth, although the fit of a sawtooth to a sinusoid cannot be very good; so probability theory will estimate the noise to be large. The same is true for harmonic frequencies with decay; however, the estimated amplitude and phase of the signal will not be accurate. This interpretation should also warn you that when you try to fit a semiperiodic signal (like a Bessel oscillation) to a single sinusoidal model, the fit will be poor. Fundamentally, the spectrum of a nonsinusoidal signal does not have a sharp peak; and so the sharpness of the periodogram is no longer a criterion for how well its parameters can be determined.

6.1.5 Nonuniform Sampling

All of the analysis done in Chapters 2 through 5 is valid when the sampling intervals are nonuniformly spaced. But is anything to be gained by using such a sampling technique? Initially we might anticipate that the problem of aliasing will be significantly reduced. Additionally, the low frequency cutoff is a function of the length of time one samples. We will be using samples of the same duration, so we do not expect to see any significant change in the ability to detect and resolve low frequencies. But will the ability to detect any signal be changed? Will sampling at a nonuniform rate make it possible to estimate a frequency better? We will attempt to address all of these concerns. But most of this will be in the form of numerical demonstrations. No complete analytical theory exists on this subject.

Aliasing

We will address the question of aliasing first. To make this test as clear as possible we have performed it without noise. The data were generated using

$$d_j = \cos([\pi + 0.3]t_j + 1).$$

For the uniform sampled data t_j is a simple index running from $-T$ to T by integer steps and $2T + 1 = 512$. Except for the lack of noise and the addition of π to the frequency this is just the example used in Fig. 6.1. Figure 6.5(A) is a plot of this uniformly sampled series. The true frequency is $0.3 + \pi$, but the plot has the appearance of a frequency of only 0.3 radians per unit step. In the terminology introduced by Tukey, this is an "alias" of the true frequency. The true frequency is oscillating more than one full cycle for each time step measured. The nonaliased frequencies that can be discriminated, with uniform time samples, have ranges from 0 to $\pi/2$. The periodogram of these data Fig. 6.5(B) has four peaks in the range $0 \leq \omega \leq 2\pi$: the true frequency at $0.3 + \pi$ and three aliases.

The nonuniform sampled time series Fig. 6.5(C) also has a time variable which takes on values from $-T$ to T. There are also 512 data points. The true frequency is unchanged. The time variable was randomly sampled. A random number generator with uniform distribution was used to generate 512 random numbers. These numbers were scaled onto the proper time intervals and the simulated signal was then evaluated at these points. No one particular region was intentionally sampled more than any

Figure 6.5: Why Aliases Exist

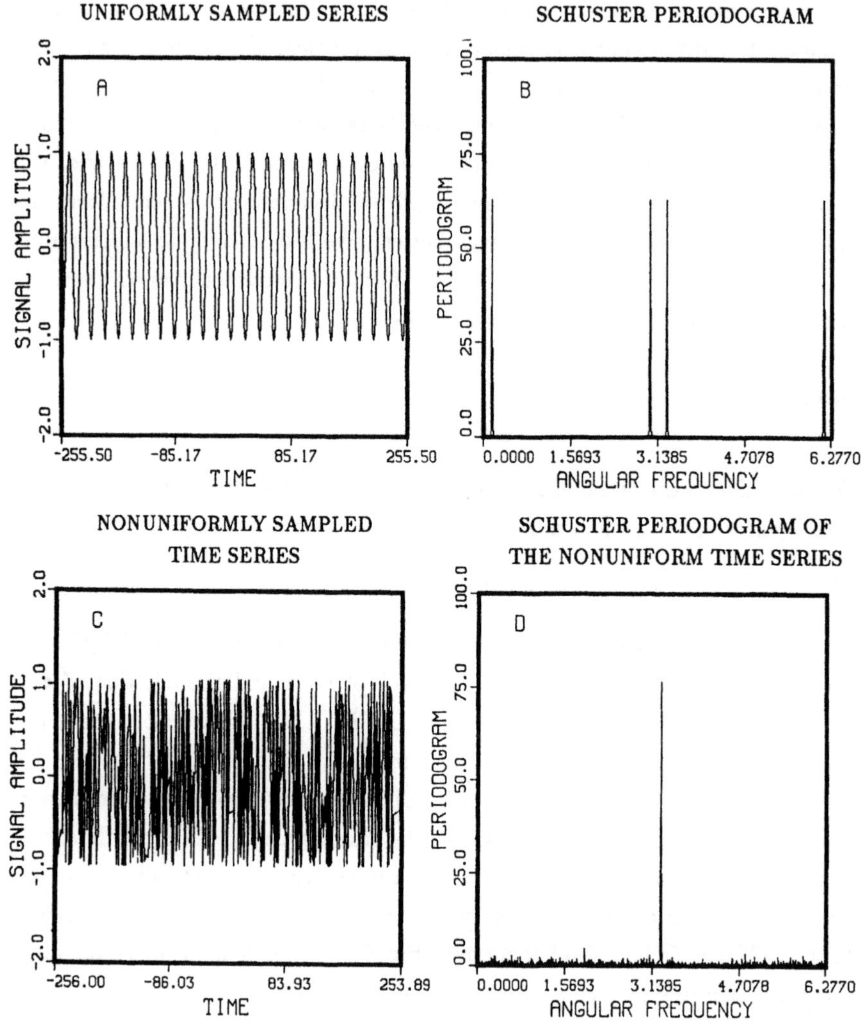

Aliasing is caused by uniform sampling of data. To demonstrate this we have prepared two sets of data: A uniformly sampled set (A), and a nonuniformly sampled set (C). The periodogram for the uniform signal, (B), contains a peak at the true frequency $(0.3 + \pi)$ plus three alias peaks. The periodogram of the nonuniformly sampled data, (D) has no aliases.

other. The time series, Fig. 6.5(C) looks very different from the uniformly sampled time series. By sampling at different intervals the presence of the high frequency becomes apparent.

We then computed the periodogram for these data and have displayed it in Fig. 6.5(D). The first striking feature is that the aliased frequencies (those corresponding to negative frequencies as well as those due to adding π to the frequency) have disappeared. The second feature which is apparent is that the two periodograms are nearly the same height. Sampling at a nonuniform rate does not significantly alter the precision of the frequency estimates, provided we have the same amount of data, and the same total sampling time. Third, the nonuniformly sampled time series has small features in the periodogram which look very much like noise, even though we know the signal has no noise. Small wiggles in the periodogram are not caused just by the noise; they can also be caused by the irregular sampling times (it should be remembered that these features are not relevant to the parameter estimation problem). The answer to the first question: "Will aliasing go away when one uses a nonuniformly sampled time series?" is yes.

Why aliasing is eliminated for a nonuniform time series is easily understood. Consider Fig. 6.6; here we have illustrated the true frequency (solid line) and the three alias frequencies from the previous example, Fig. 6.5. The squares mark the location of three uniform sample points, while the circles mark the location of the nonuniform points. Looking at Fig. 6.6 we now see aliasing in an entirely different light. Probability theory is indicating (quite rightly) that in the frequency region $0 \leq \omega \leq 2\pi$ there are four equally probable frequencies, Fig. 6.5(B), while for the nonuniformly sampled data, probability theory is indicating that there is only one frequency consistent with the data, Fig. 6.5(D).

Of course it must be true that the aliasing phenomenon returns for some sufficiently high frequency. If the sampling times $\{t_i\}$, although nonuniform, are all integer multiples of some small interval δt, then frequencies differing by $2\pi/\delta t$ will still be aliased. We did one numerical test with a signal-to-noise ratio of one and the same true frequency. We then calculated the periodogram for higher frequencies. We continued increasing the frequency until we obtained the first alias. This occurred at a frequency around 60π, almost 1.8 orders of magnitude improvement in the frequency band free of aliasing. Even then this second large maximum was many orders of magnitude below that at the first "true" frequency.

Figure 6.6: Why Aliases Go Away for Nonuniformly Sampled Data

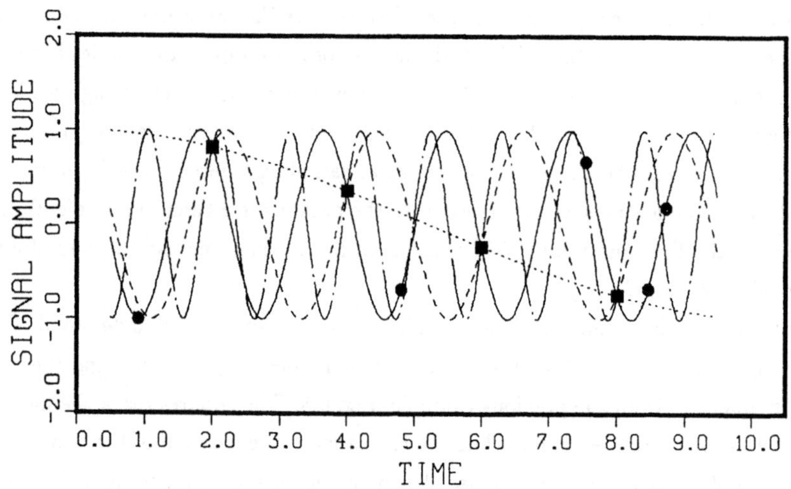

When aliases occur in a uniformly sampled time series, probability theory is still working correctly; indicating there is more than one frequency corresponding to the "best" estimate in the data. Suppose we have a signal $\cos(0.3 + \pi)t$ (solid line). For a uniformly sampled data (squares) there are four possible frequencies: $\hat{\omega} = 0.3$ (dotted line), $\hat{\omega} = \pi - 0.3$ (dashed line), $\hat{\omega} = 2\pi - 0.3$ (chain dot), and the true frequency (solid line) which pass through the uniform data (marked with squares). For nonuniformly sampled time series (marked with circles), aliases effectively do not occur because only the "true" (solid line) signal passes through the nonuniformly spaced points (circles).

Figure 6.7: Uniform Sampling Compared to Nonuniform Sampling

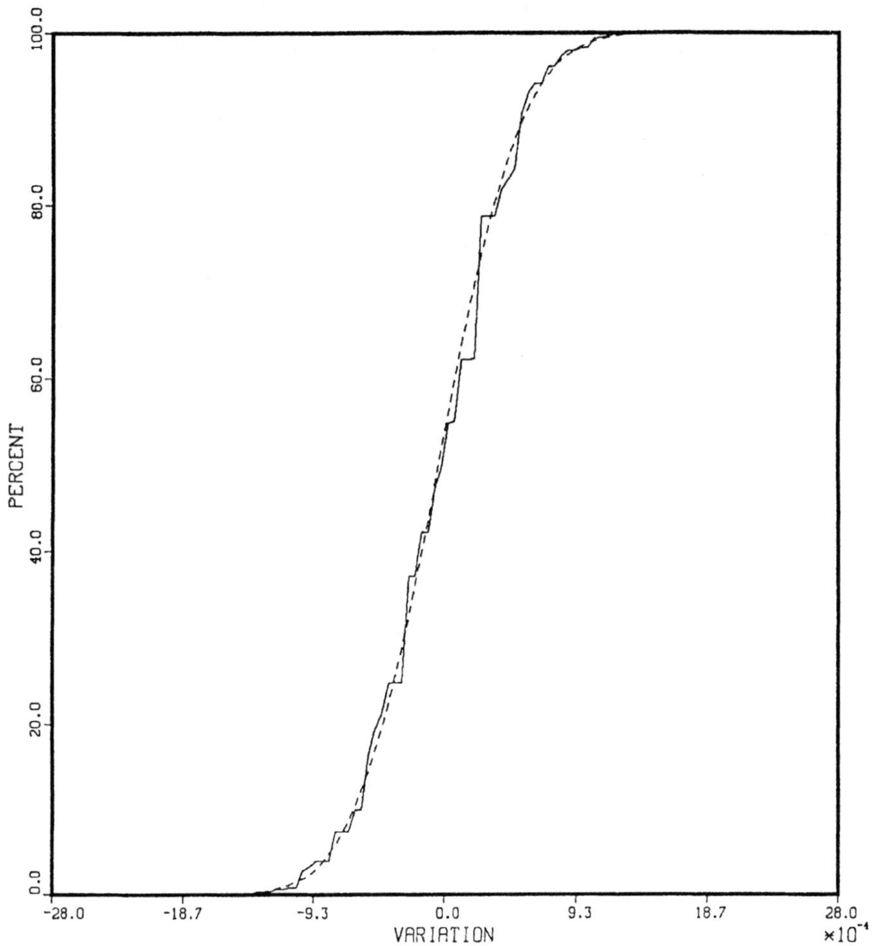

We generated some 3000 sets of data with nonuniform data samples, estimated the frequency, computed a histogram, and computed the cumulative number of estimates summing from left to right on this plot (solid line). The equivalent plot for uniformly sampled data is repeated here for easy reference (dotted line). Clearly there is no significant difference in these plots.

Nonuniform Sampling and the Frequency Estimates

The second question is "Will sampling at a nonuniform rate significantly change the frequency estimate?". To answer this question we have set up a second test, Fig. 6.7. The simulated signal is the same as that in Fig. 6.2, only now the samples are nonuniform. We generated some 3000 samples of the data and estimated the frequency from each. We then computed a histogram and integrated to obtain cumulative sampling distribution of the estimates, Fig. 6.7. If nonuniform sampling improves the frequency resolution then we would expect the cumulative distribution (solid line) to rise faster than for the uniformly sampled case (dotted line). As one can see from this plot, nonuniform sampling is clearly equivalent to uniform sampling when it comes to the accuracy of the parameter estimates; moreover, nonuniform sampling improves the high frequency resolution but does not change the frequency estimates otherwise.

Some might be disturbed by the irregular appearance of the solid line in Fig. 6.7. This irregular behavior is simply "digitization" error in the calculation. When we performed this calculation for the uniform case the first time, this same effect was present. We were unsure of the cause, so we repeated the calculation forcing our searching routines to find the maximum of the periodogram much more precisely. The irregular behavior was much reduced. We did not repeat this procedure on the nonuniformly sampled data, because it is very expensive computationally.

6.2 A Frequency with Lorentzian Decay

The simple harmonic frequency problem discussed in Chapter 2 may be generalized easily to include Lorentzian or Gaussian decay. We assume, for this discussion, that the decay is Lorentzian; the generalization to other types of decay will become more obvious as we proceed. For a uniformly sampled interval the model we are considering is

$$f(l) = [B_1 \cos(\omega l) + B_2 \sin(\omega l)]e^{-\alpha l} \tag{6.3}$$

where l is restricted to values $(1 \leq l \leq N)$. We now have four parameters to estimate: the amplitudes B_1, B_2; the frequency ω; and the decay rate α.

6.2.1 The "Student t-Distribution"

The solution to this problem is a straightforward application of the general procedures. The matrix g_{ij} (3.4) is given by

$$g_{ij} = \begin{pmatrix} \sum_{l=1}^{N} \cos^2(\omega l)e^{-2\alpha l} & \sum_{l=1}^{N} \cos\omega l \sin\omega l e^{-2\alpha l} \\ \sum_{l=1}^{N} \cos\omega l \sin\omega l e^{-2\alpha l} & \sum_{l=1}^{N} \sin^2(\omega l)e^{-2\alpha l} \end{pmatrix}.$$

This problem can be solved exactly. However, the exact solution is tedious, and not very informative. Fortunately an approximate solution is easily obtained which exhibits most of the important features of the full solution; and is valid in the same sense that a discrete Fourier transform is valid. We approximate g_{ij} as follows: First the sum over the sine squared and cosine squared terms may be approximated as

$$\begin{aligned} c &\equiv \sum_{l=1}^{N} \cos^2(\omega l)e^{-2l\alpha} \approx \sum_{l=1}^{N} \sin^2(\omega l)e^{-2l\alpha} \\ &= \frac{1}{2}\sum_{l=1}^{N}[1 \pm \cos(2\omega l)]e^{-2l\alpha} \\ &\approx \frac{1}{2}\sum_{l=1}^{N} e^{-2l\alpha} \\ &= \frac{1}{2}\left[\frac{1-e^{-2N\alpha}}{e^{2\alpha}-1}\right]. \end{aligned} \tag{6.4}$$

Second, the off diagonal terms are at most the same order as the ignored terms; these terms are therefore ignored. Thus the matrix g_{ij} can be approximated as

$$g_{ij} \approx \begin{pmatrix} c & 0 \\ 0 & c \end{pmatrix}.$$

The orthonormal model functions may then be written as

$$H_1(l) = \cos(\omega l)e^{-\alpha l}/\sqrt{c} \tag{6.5}$$

$$H_2(l) = \sin(\omega l)e^{-\alpha l}/\sqrt{c} \tag{6.6}$$

The projections of the data onto the orthonormal model functions (3.13) are given by

$$h_1 \equiv \frac{R(\omega,\alpha)}{\sqrt{c}} = \frac{1}{\sqrt{c}}\sum_{l=1}^{N} d_l \cos(\omega l)e^{-\alpha l}$$

$$h_2 \equiv \frac{I(\omega,\alpha)}{\sqrt{c}} = \frac{1}{\sqrt{c}}\sum_{l=1}^{N} d_l \sin(\omega l)e^{-\alpha l}$$

and the joint posterior probability of a frequency ω and a decay rate α is given by

$$P(\omega,\alpha|D,I) \propto \left[1 - \frac{R(\omega,\alpha)^2 + I(\omega,\alpha)^2}{N c\overline{d^2}}\right]^{\frac{2-N}{2}}. \quad (6.7)$$

This approximation is valid provided there are plenty of data, $N \gg 1$, and there is no evidence of a low frequency. There is no restriction on the range of α: if $\alpha > 0$ the signal is decaying with increasing time, if $\alpha < 0$ the signal is growing with increasing time, and if $\alpha = 0$ the signal is stationary. This equation is analogous to (2.8) and reduces to (2.8) in the limit $\alpha \to 0$.

6.2.2 Accuracy Estimates

We derived a general estimate for the $\{\omega\}$ parameters in Chapter 4, and we would like to use those estimates for comparison with the single stationary frequency problem. To do this we can approximate the probability distribution $P(\omega,\alpha|D,\sigma,I)$ by a Gaussian as was done in Chapter 4. This may be done readily by assuming a form of the data, and then applying Eqs. (4.9) through (4.14). From the second derivative we may obtain the desired (mean) $\pm$ (standard deviation) estimates. Approximate second derivatives, usable with real data, may be obtained analytically as follows. We take as the data

$$d(t) = \hat{B}\cos(\hat{\omega}t)e^{-\hat{\alpha}t}, \quad (6.8)$$

where $\hat{\omega}$ is the true frequency of oscillation and $\hat{\alpha}$ is the true decay rate. We have assumed only a cosine component to effect some simplifications in the discussion. It will be obvious at the end of the calculation that the result for a signal of arbitrary phase and magnitude may be obtained by replacing the amplitude $\hat{B}^2$ by the squared magnitude $\hat{B}^2 \to \hat{B}_1^2 + \hat{B}_2^2$.

The projection of the data (6.8) onto the model functions (6.5), and (6.6) is:

$$h_1 = \frac{\hat{B}}{2\sqrt{c}}\left[\sum_{l=1}^{N}\cos(\omega-\hat{\omega})l e^{-(\alpha+\hat{\alpha})l} + \sum_{l=1}^{N}\cos(\omega+\hat{\omega})l e^{-(\alpha+\hat{\alpha})l}\right].$$

The second term is negligible compared to the first under the conditions we have in mind. Likewise, the projection h_2 is essentially zero compared to h_1 and we have ignored it. These sums may be done explicitly using (6.4) to obtain

$$h_1 = \frac{\hat{B}}{4\sqrt{c}}\left[\frac{1-e^{-2Nv}}{e^{2v}-1}+\frac{1-e^{-2Nu}}{e^{2u}-1}\right]$$

where

$$v = \frac{\alpha+\hat{\alpha}-i(\omega-\hat{\omega})}{2} \qquad \text{and} \qquad u = \frac{\alpha+\hat{\alpha}+i(\omega-\hat{\omega})}{2},$$

and $i=\sqrt{-1}$ in the above equations. Then the sufficient statistic $\overline{h^2}$ is given by:

$$\overline{h^2} = \frac{\hat{B}^2}{16}\left[\frac{e^{2\alpha}-1}{1-e^{-2N\alpha}}\right]\left[\frac{1-e^{-2Nv}}{1-e^{2v}}+\frac{1-e^{-2Nu}}{1-e^{2u}}\right]^2$$

The region of the parameter space we are interested in is where the unitless decay rate is small compared to one, and $\exp(N\hat{\alpha})$ is large compared to one. In this region the true signal decays away in the observation time, but not before we obtain a good representative sample of it. We are not considering the case were the decay is so slow that the signal is nearly stationary, or so fast that the signal is gone within a small fraction of the observation time. Within these limits the sufficient statistic $\overline{h^2}$ is

$$\overline{h^2} \approx \frac{\hat{B}^2\alpha}{2}\left[\frac{\alpha+\hat{\alpha}}{(\alpha+\hat{\alpha})^2+(\omega-\hat{\omega})^2}\right]^2.$$

The first derivatives of $\overline{h^2}$ evaluated at $\omega=\hat{\omega}$ and $\alpha=\hat{\alpha}$ are zero, as they should be. The mixed second partial derivative is also zero, indicating that the presence of decay does not (to second order) shift the location of a frequency, and this of course explains why the discrete Fourier transform works on problems with decay. This gives the second derivatives of $\overline{h^2}$ as

$$b_\alpha \equiv -\left(\frac{\partial^2\overline{h^2}}{\partial\alpha^2}\right)_{\alpha=\hat{\alpha}} = \frac{\hat{B}^2}{16\hat{\alpha}^3} \qquad \text{and} \qquad b_\omega \equiv -\left(\frac{\partial^2\overline{h^2}}{\partial\omega^2}\right)_{\omega=\hat{\omega}} = \frac{\hat{B}^2}{2\hat{\alpha}^3}.$$

From these derivatives we then make the (mean) $\pm$ (standard deviation) estimates of the frequency and decay rate to obtain

$$(\alpha)_{\text{est}} = \hat{\alpha} \pm \frac{\sigma}{\sqrt{b_\alpha}} \qquad \text{and} \qquad (\omega)_{\text{est}} = \hat{\omega} \pm \frac{\sigma}{\sqrt{b_\omega}}$$

where

$$\frac{\sigma}{\sqrt{b_\alpha}} \approx \frac{2.8\sigma\hat{\alpha}^{\frac{3}{2}}}{|\hat{B}|} \qquad \text{and} \qquad \frac{\sigma}{\sqrt{b_\omega}} \approx \frac{\sigma\hat{\alpha}^{\frac{3}{2}}}{|\hat{B}|}. \tag{6.9}$$

Converting to physical units, if the sampling rate is Δt and $\hat{\alpha}$ is now the true decay rate in hertz, these accuracy estimates are

$$\frac{\sigma}{\sqrt{b_\alpha}} \approx \frac{2.8\sigma}{|\hat{B}|}\sqrt{\hat{\alpha}^3\Delta t} \text{ Hertz} \qquad \text{and} \qquad \frac{\sigma}{\sqrt{b_\omega}} \approx \frac{\sigma}{|\hat{B}|}\sqrt{\hat{\alpha}^3\Delta t} \text{ Hertz.}$$

Just as with the single frequency problem the accuracy depends on the signal-to-noise ratio and on the amount of data. In the single frequency case the amount of data was represented by the factor of N. Here the amount of data depends on two factors: the true decay rate $\hat{\alpha}$, and the sampling time Δt. The only factor the experimenter can typically control is the sampling time Δt. With a decaying signal, to improve the accuracy of the parameter estimates one must take the data faster, thus ensuring that the data are sampled in the region where the signal is large, or one must improve the signal-to-noise ratio of the data.

How does this compare to the results obtained before for the simple harmonic frequency? For a signal with $N = 1000$, a decay rate of $\hat{\alpha} = 2$Hz, $\hat{B}/\sqrt{2}\sigma = 1$, and again taking data for 1 second gives the accuracy estimates for frequency and decay as

$$(\omega)_{\text{est}} = \hat{\omega} \pm 0.06 \text{ Hz} \qquad \text{and} \quad (\alpha)_{\text{est}} = \hat{\alpha} \pm 0.17 \text{ Hz}.$$

The uncertainty in ω is 0.13Hz compared to 0.025Hz for an equivalent stationary signal with the same signal-to-noise ratio. This is a factor of 2.4 times larger than for a stationary sinusoid, and since the error varies like $N^{-\frac{3}{2}}$ we have effectively lost all but one third of the data due to decay. When we have reached the unitless time of $t = 250$ the signal is down by a factor of 12 and has all but disappeared into the noise. Again, the results of probability theory correspond nicely to the indications of common sense – but they are quantitative where common sense is not.

6.2.3 Example – One Frequency with Decay

To illustrate some of these points we been making, we have prepared two more examples: first we will investigate the use of this probability density when the decay mode is known and second when it is unknown.

Known method of decay

Figure 6.8 is an example of the use of the posterior probability (6.7) when the signal is known to be harmonic with Lorentzian decay. This time series was prepared from the following equation

$$d_j = 0.001 + \cos(0.3j + 1)e^{-0.01j} + e_j. \tag{6.10}$$

The $N = 512$ data samples were prepared in the following manner: first, we generated the data without the noise; we then computed the average of the data, and subtracted

Figure 6.8: Single Frequency with Lorentzian Decay

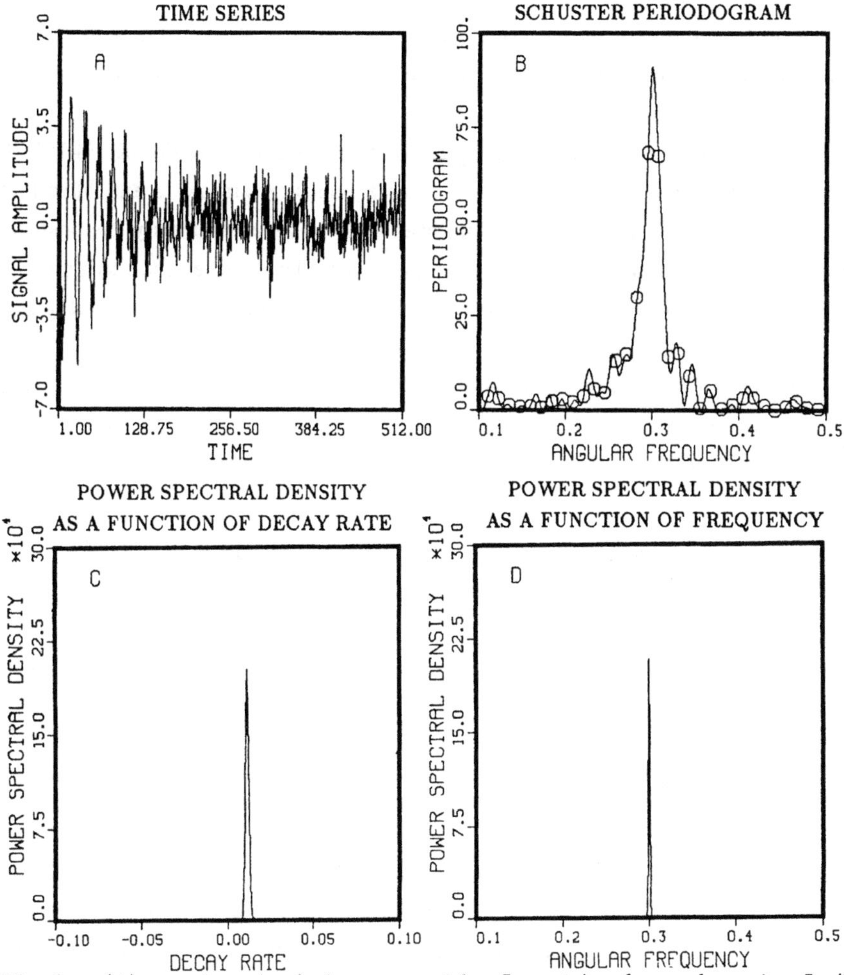

The data (A) contain a simple frequency with a Lorentzian decay plus noise. In (B) the noise has significantly distorted the periodogram (continuous curve) and the fast Fourier transform (open circles). The power spectral density may be computed as a function of decay rate α by integrating over the frequency (C), or as a function of frequency ω by integrating over the decay (D).

it from each data point, to ensure that the average of the data is zero; we then repeated this process on the Gaussian white noise; next, we scaled the computer generated signal by the appropriate ratio to make the signal-to-noise ratio of the data analyzed exactly one. The time series clearly shows a small signal which rapidly decays away, Fig. 6.8(A). Figure 6.8(B), the periodogram (continuous curve) and the fast Fourier transform (open circles) clearly show the Lorentzian line shape. The noise is now significantly affecting the periodogram: the periodogram is no longer an optimum frequency estimator.

Figures 6.8(C) and 6.8(D) contain plots of the power spectral density (4.16). In Fig. 6.8(C) we have treated the frequency as a nuisance parameter and have integrated it out numerically; as was emphasized earlier this is essentially the posterior probability distribution for α normalized to a power level rather than to unity. In Fig. 6.8(D) we have treated the decay as the nuisance parameter and have integrated it out. This gives the power spectral estimate as a function of frequency.

The width of these curves is a measure of the uncertainty in the determination of the parameters. We have determined full-width at half maximum (numerically) for each of these and have compared these to the theoretical "best" estimates (6.9) and find

$$(\omega)_{\text{est}} = 0.2998 \pm 5 \times 10^{-4} \qquad \text{and} \qquad (\omega)_{\text{best}} = 0.3000 \pm 6 \times 10^{-4},$$

$$(\alpha)_{\text{est}} = 0.0109 \pm 1.6 \times 10^{-3} \qquad \text{and} \qquad (\alpha)_{\text{best}} = 0.0100 \pm 1.6 \times 10^{-3}.$$

The theoretical estimates and those calculated from these data are effectively identical.

Unknown Method of Decay

Now what effect does not knowing the true model have on the estimated accuracy of these parameters? To test this we have analyzed the signal from Fig. 6.8 using four different models and have summarized the results in Table . There are several significant observations about the accuracy estimates; including a decay mode does not significantly affect the frequency estimates; however it does improve the accuracy estimates for the frequency as well as the estimated standard deviation of the noise σ, but not very much.

As we had expected, the Gaussian decay does not fit the data well: it decays away too fast, and the accuracy estimates are a little poorer. As with the single

Table 6.1: The Effect of Not Knowing the Decay Mode

Description	model	frequency ω	σ	$P(f_j\|D,I)$
Stationary:	$B\cos(\omega t+\theta)$	$0.3001\pm 6\times 10^{-4}$	1.260	8.3×10^{-33}
Gaussian in time:	$B\cos(\omega t+\theta)e^{-\alpha t^2}$	$0.2991\pm 7\times 10^{-4}$	0.993	6.5×10^{-4}
Lorentzian in time:	$\dfrac{B\cos(\omega t+\theta)}{1+\alpha t^2}$	$0.2998\pm 5\times 10^{-4}$	0.978	0.0027
Lorentzian in frequency:	$B\cos(\omega t+\theta)e^{-\alpha t}$	$0.2998\pm 5\times 10^{-4}$	0.979	0.9972

We analyzed the single frequency plus decay data (6.10) using four different decay models: stationary harmonic frequency, Gaussian decay, Lorentzian in time, and last Lorentzian in frequency. The stationary harmonic frequency model (first row) gives a poor estimate of the standard deviation of the noise, and consequently the estimated uncertainty of the frequency is larger. The probability of this model is so small that one would not even consider this as a possible model of the data. The second model is a single frequency with Gaussian decay. Here the estimated standard deviation of the noise is accurate, but the model fits the data poorly; thus the relative probability of this model effectively eliminates it from consideration. The third model is a single frequency with a Lorentzian decay in time. The relative probability of this model is also small indicating that although it is better than the two previous models, it is not nearly as good as the last model. The last model is a single frequency with Lorentzian decay. The relative probability of the model is effectively one, within the class of models considered.

harmonic frequency problem when we were demonstrating the effects of violating the assumptions, nothing startling happens here and maybe that is the most startling thing of all. Because it means that we do not have to know the exact models to make significant progress on analyzing the data. All we need are models which are reasonable for the data; i.e. models which take on most of the characteristics of the data.

The last column in this table is the relative probability of the various models (5.1). The relative probability of the single harmonic frequency model, 8.3×10^{-33} completely rules this model out as a possible explanation of these data. This is again not surprising: one can look at the data and see that it is decaying away. This small probability is just a quantitative way of stating a conclusion that we draw so easily without any probability theory. The Gaussian model fits the data much better, 6.5×10^{-4}, but not as well as the two Lorentzian models. The Lorentzian model in time has only about one chance in 500 of being "right" (i.e. of providing a better description of future data than the Lorentzian in frequency). Thus probability theory can rank various models according to how well they fit the data, and discriminates easily between models which predict only slightly different data.

6.3 Two Harmonic Frequencies

We now turn our attention to the slightly more general problem of analyzing a data set which we postulate contains two distinct harmonic frequencies. The "Student t-distribution" represented by (3.17) is, of course, the general solution to this problem. Unfortunately, that equation does not lend itself readily to understanding intuitively what is in the probability distribution. In particular we would like to know the behavior of these equations in three different limits: first, when the frequencies are well separated; second, when they are close but distinct; and third, when they are so close as to be, for all practical purposes, identical. To investigate these we will solve, approximately, the two stationary frequency problem.

6.3.1 The "Student t-Distribution"

The model equation for the two-frequency problem is a simple generalization of the single-harmonic problem:

$$f(t) = B_1\cos(\omega_1 t) + B_2\cos(\omega_2 t) + B_3\sin(\omega_1 t) + B_4\sin(\omega_2 t).$$

The model functions can then be used to construct the g_{jk} matrix. On a uniform grid this is given by

$$g_{jk} = \begin{pmatrix} c_{11} & c_{12} & 0 & 0 \\ c_{12} & c_{22} & 0 & 0 \\ 0 & 0 & s_{11} & s_{12} \\ 0 & 0 & s_{12} & s_{22} \end{pmatrix}$$

where

$$c_{jk} = \sum_{l=-T}^{T} \cos(\omega_j l)\cos(\omega_k l) = \frac{\sin(\frac{1}{2}N\omega_+)}{2\sin(\frac{1}{2}\omega_+)} + \frac{\sin(\frac{1}{2}N\omega_-)}{2\sin(\frac{1}{2}\omega_-)} \tag{6.11}$$

$$s_{jk} = \sum_{l=-T}^{T} \sin(\omega_j l)\sin(\omega_k l) = \frac{\sin(\frac{1}{2}N\omega_-)}{2\sin(\frac{1}{2}\omega_-)} - \frac{\sin(\frac{1}{2}N\omega_+)}{2\sin(\frac{1}{2}\omega_+)} \tag{6.12}$$

$$\omega_+ = \omega_j + \omega_k, \qquad (j,k = 1 \text{ or } 2)$$

$$\omega_- = \omega_j - \omega_k.$$

The eigenvalue and eigenvector problem for g_{jk} splits into two separate problems, each involving 2×2 matrices. The eigenvalues are:

$$\lambda_1 = \frac{c_{11}+c_{22}}{2} + \sqrt{(c_{11}-c_{22})^2 + 4c_{12}^2}, \qquad \lambda_2 = \frac{c_{11}+c_{22}}{2} - \sqrt{(c_{11}-c_{22})^2 + 4c_{12}^2},$$

$$\lambda_3 = \frac{s_{11}+s_{22}}{2} + \sqrt{(s_{11}-s_{22})^2 + 4s_{12}^2}, \quad \text{and} \quad \lambda_4 = \frac{s_{11}+s_{22}}{2} - \sqrt{(s_{11}-s_{22})^2 + 4s_{12}^2}.$$

Well Separated Frequencies

When the frequencies are well separated $|\omega_1 - \omega_2| \gg 2\pi/N$, the eigenvalues reduce to $\lambda = N/2$. That is, g_{jk} goes into $N/2$ times the unit matrix. Then the model functions are effectively orthogonal and the sufficient statistic $\overline{h^2}$ reduces to

$$\overline{h^2} = \frac{2}{N}\left[C(\omega_1) + C(\omega_2)\right].$$

The joint posterior probability, when the variance is known, is given by

$$P(\omega_1, \omega_2|D, \sigma, I) \propto \exp\left[\frac{C(\omega_1) + C(\omega_2)}{\sigma^2}\right]. \tag{6.13}$$

The problem has separated: one can estimate each of the frequencies separately. The maximum of the two-frequency posterior probability density will be located at the two greatest peaks in the periodogram, in agreement with the common sense usage of the discrete Fourier transform.

Two Very Close Frequencies

The labels ω_1, ω_2, etc. for the frequencies in the model are arbitrary, and accordingly their joint probability density is invariant under permutations. That means, for the two-frequency problem, there is an axis of symmetry running along the line $\omega_1 = \omega_2$. We do not know from (6.13) what is happening along that line. This is easily investigated: when $\omega_1 = \omega_2 \equiv \omega$ the eigenvalues become

$$\lambda_1 = N, \qquad \lambda_2 = 0, \qquad \lambda_3 = N, \qquad \lambda_4 = 0.$$

The matrix g_{jk} has two redundant eigenvalues, and the probability distribution becomes

$$P(\omega|D,\sigma,I) \propto \exp\left\{\frac{C(\omega)}{\sigma^2}\right\}. \tag{6.14}$$

The probability density goes smoothly into the single frequency probability distribution along this axis of symmetry. Given that the two frequencies are equal, our estimate of them will be identical, in value and accuracy, to those of the one frequency case. In the exact solution, the factor of two that we would have if we attempted to use (6.13) where it is not valid, is just cancelled out.

Close But Distinct Frequencies

We have not yet addressed the posterior probability density when there are two close but distinct frequencies. To understand this aspect of the problem we could readily diagonalize the matrix g_{jk} and obtain the exact solution. However, just like the single frequency case with Lorentzian decay, this would be extremely tedious and not very productive. Instead we derive an approximate solution which is simpler and valid nearly everywhere if N is large. To obtain this approximate solution one needs only to examine the matrix g_{jk} and notice that the elements of this matrix consist of the diagonal elements given by:

$$c_{11} = \frac{N}{2} + \frac{\sin(N\omega_1)}{2\sin(\omega_1)} \approx \frac{N}{2},$$

$$c_{22} = \frac{N}{2} + \frac{\sin(N\omega_2)}{2\sin(\omega_2)} \approx \frac{N}{2},$$

$$s_{11} = \frac{N}{2} - \frac{\sin(N\omega_1)}{2\sin(\omega_1)} \approx \frac{N}{2},$$

$$s_{22} = \frac{N}{2} - \frac{\sin(N\omega_2)}{2\sin(\omega_2)} \approx \frac{N}{2},$$

and the off-diagonal elements. The off-diagonal terms are small compared to N unless the frequencies are specifically in the region of $\omega_1 \approx \omega_2$; then only the terms involving the difference $(\omega_1 - \omega_2)$ are large. We can approximate the off diagonal terms as:

$$c_{12} \approx s_{12} \approx \frac{1}{2}\sum_{l=-T}^{T} \cos\frac{1}{2}(\omega_1 - \omega_2)l = \frac{1}{2}\frac{\sin\frac{1}{2}N(\omega_1 - \omega_2)}{\sin\frac{1}{2}(\omega_1 - \omega_2)} \equiv \frac{B}{2}. \tag{6.15}$$

When the two frequencies are well separated, (6.15) is of order one and is small compared to the diagonal elements. When the two frequencies are nearly equal, then the off-diagonal terms are large and are given accurately by (6.15). So the approximation is valid for all values of ω_1 and ω_2 that are not extremely close to zero and π.

With this approximation for g_{jk} it is now possible to write a simplified solution for the two-frequency problem. The matrix g_{jk} is given approximately by

$$g_{jk} = \frac{1}{2}\begin{pmatrix} N & B & 0 & 0 \\ B & N & 0 & 0 \\ 0 & 0 & N & B \\ 0 & 0 & B & N \end{pmatrix}.$$

The orthonormal model functions (3.5) may now be constructed:

$$H_1(t) = \frac{1}{\sqrt{N+B}}\left[\cos(\omega_1 t) + \cos(\omega_2 t)\right], \tag{6.16}$$

$$H_2(t) = \frac{1}{\sqrt{N-B}}\left[\cos(\omega_1 t) - \cos(\omega_2 t)\right],$$

$$H_3(t) = \frac{1}{\sqrt{N+B}}\left[\sin(\omega_1 t) + \sin(\omega_2 t)\right],$$

$$H_4(t) = \frac{1}{\sqrt{N-B}}\left[\sin(\omega_1 t) - \sin(\omega_2 t)\right].$$

We can write the sufficient statistic $\overline{h^2}$ in terms of these orthonormal model functions to obtain

$$\overline{h^2} = h_+^2 + h_-^2,$$

$$h_+^2 \equiv \frac{1}{4(N+B)}\left\{[R(\omega_1) + R(\omega_2)]^2 + [I(\omega_1) + I(\omega_2)]^2\right\},$$

$$h_-^2 \equiv \frac{1}{4(N-B)}\left\{[R(\omega_1) - R(\omega_2)]^2 + [I(\omega_1) - I(\omega_2)]^2\right\},$$

where R and I are the sine and cosine transforms of the data as functions of the appropriate frequency. The factor of 4 comes about because for this problem there are $m = 4$ model functions. Using (3.15), the posterior probability that two distinct frequencies are present given the noise variance σ^2 is

$$P(\omega_1, \omega_2|D, \sigma, I) \propto \exp\left\{\frac{2\overline{h^2}}{\sigma^2}\right\}. \tag{6.17}$$

A quick check on the asymptotic forms of this will verify that when the frequencies are well separated one has $\overline{h^2} = \frac{1}{2}[C(\omega_1) + C(\omega_2)]$, and it has reduced to (6.13). Likewise, when the frequencies are the same the second term goes smoothly to zero, and the first term goes into $\frac{1}{2}C(\omega)$, to reduce to (6.14) as expected.

6.3.2 Accuracy Estimates

When the frequencies are very close or far apart we can apply the results obtained by Jaynes [12] concerning the accuracy of the frequency estimates:

$$(\omega)_{\text{est}} = \hat{\omega} \pm \frac{\sigma}{\hat{B}}\sqrt{48/N^3}. \tag{6.18}$$

In the region where the frequencies are close but distinct, (6.17) appears very different. We would like to understand what is happening in this region, in particular we would like to know just how well two close frequencies can be estimated. To understand this we will construct a Gaussian approximation similar to what was done for the case with Lorentzian decay. We Taylor expand the $\overline{h^2}$ in (6.17) to obtain

$$P(\omega_1, \omega_2|D, \sigma, I) \approx \exp\left\{-\frac{1}{2\sigma^2}\sum_{j=1}^{2}\sum_{k=1}^{2} b_{jk}(\omega_j - \hat{\omega}_j)(\omega_k - \hat{\omega}_k)\right\}$$

where

$$\begin{aligned} b_{11} &= -2\left.\frac{\partial^2 \overline{h^2}}{\partial \omega_1^2}\right|_{\substack{\omega_1=\hat{\omega}_1 \\ \omega_2=\hat{\omega}_2}} \\ b_{22} &= -2\left.\frac{\partial^2 \overline{h^2}}{\partial \omega_2^2}\right|_{\substack{\omega_1=\hat{\omega}_1 \\ \omega_2=\hat{\omega}_2}} \\ b_{12} &= -2\left.\frac{\partial^2 \overline{h^2}}{\partial \omega_1 \partial \omega_2}\right|_{\substack{\omega_1=\hat{\omega}_1 \\ \omega_2=\hat{\omega}_2}} \end{aligned}$$

where $\hat{\omega}_1$, $\hat{\omega}_2$ are the locations of the maxima of (6.17). If we have uniformly sampled data of the form

$$d_l = \hat{A}_1 \cos(\hat{\omega}_1 l) + \hat{A}_2 \cos(\hat{\omega}_2 l) + \hat{A}_3 \sin(\hat{\omega}_1 l) + \hat{A}_4 \sin(\hat{\omega}_2 l) \tag{6.19}$$

where $-T \leq l \leq T$, $2T+1 = N$, $\hat{A}_1$, $\hat{A}_2$, $\hat{A}_3$, $\hat{A}_4$ are the true amplitudes, and $\hat{\omega}_1$, $\hat{\omega}_2$ are the true frequencies, then h_j is given by the projection of H_j (6.16) onto the data (6.19) to obtain

$$h_j = \frac{1}{\sqrt{N + B(\omega_1, \omega_2)}} \sum_{l=-T}^{T} H_j(t_l) d_l$$

where

$$\frac{B(\omega_1, \omega_2)}{2} \equiv \frac{1}{2} \sum_{l=-T}^{T} \cos(\omega_1 - \omega_2) l = \frac{1}{2} \frac{\sin \frac{1}{2} N(\omega_1 - \omega_2)}{\sin \frac{1}{2}(\omega_1 - \omega_2)}. \tag{6.20}$$

For a uniform time series these h_j may be summed explicitly using (6.20) to obtain

$$h_1 = \frac{1}{2\sqrt{N + B(\omega_1, \omega_2)}} \times \Big\{ \hat{A}_1 [B(\hat{\omega}_1, \omega_1) + B(\hat{\omega}_1, \omega_2)] + \hat{A}_2 [B(\hat{\omega}_2, \omega_1) + B(\hat{\omega}_2, \omega_2)] \Big\}$$

$$h_2 = \frac{1}{2\sqrt{N - B(\omega_1, \omega_2)}} \times \Big\{ \hat{A}_1 [B(\hat{\omega}_1, \omega_1) - B(\hat{\omega}_1, \omega_2)] + \hat{A}_2 [B(\hat{\omega}_2, \omega_1) - B(\hat{\omega}_2, \omega_2)] \Big\}$$

$$h_3 = \frac{1}{2\sqrt{N + B(\omega_1, \omega_2)}} \times \Big\{ \hat{A}_3 [B(\hat{\omega}_1, \omega_1) + B(\hat{\omega}_1, \omega_2)] + \hat{A}_4 [B(\hat{\omega}_2, \omega_1) + B(\hat{\omega}_2, \omega_2)] \Big\}$$

$$h_4 = \frac{1}{2\sqrt{N - B(\omega_1, \omega_2)}} \times \Big\{ \hat{A}_3 [B(\hat{\omega}_1, \omega_1) - B(\hat{\omega}_1, \omega_2)] + \hat{A}_4 [B(\hat{\omega}_2, \omega_1) - B(\hat{\omega}_2, \omega_2)] \Big\}.$$

We have kept terms corresponding to the differences in the frequencies. When the frequencies are close together it is only these terms which are important: the approximation is consistent with the others made.

The sufficient statistic $\overline{h^2}$ is then given by

$$\overline{h^2} = \frac{1}{4}(h_1^2 + h_2^2 + h_3^2 + h_4^2). \tag{6.21}$$

To obtain a Gaussian approximation for (6.17) one must calculate the second derivative of (6.21) with respect to ω_1 and ω_2. The problem is simple in principle but tedious in practice. To get these partial derivatives, we Taylor expand (6.21) around the maximum located at $\hat{\omega}_1$ and $\hat{\omega}_2$ and then take the derivative. The intermediate steps are of little concern and were carried out using an algebra manipulation package. Terms of order one compared to N were again ignored, and we have assumed the frequencies are close but distinct, also we used the small angle approximations for the sine and cosine at the end of the calculation. The local variable δ [defined as $(\hat{\omega}_2 - \hat{\omega}_1)/2 \equiv \delta/N$] measures the distance between two adjacent frequencies. If δ is π then the frequencies are separated by one step in the discrete Fourier transform. The second partial derivatives of $\overline{h^2}$ evaluated at the maximum are given by:

$$\begin{aligned} b_{11} &\approx (\hat{A}_1^2 + \hat{A}_3^2)N^3 \left(\frac{3\sin^2\delta - 6\delta\cos\delta\sin\delta + \delta^2[\sin^2\delta + 3\cos\delta] - \delta^4}{24\delta^3[\sin\delta - \delta][\sin\delta + \delta]}\right) \\ b_{22} &\approx (\hat{A}_2^2 + \hat{A}_4^2)N^3 \left(\frac{3\sin^2\delta - 6\delta\cos\delta\sin\delta + \delta^2[\sin^2\delta + 3\cos\delta] - \delta^4}{24\delta^3[\sin\delta - \delta][\sin\delta + \delta]}\right) \\ b_{12} &\approx (\hat{A}_1\hat{A}_2 + \hat{A}_3\hat{A}_4)N^3 \left(\frac{\delta^4\sin\delta + 2\delta^3\cos\delta - 3\delta^2\sin\delta + \sin^3\delta}{8\delta^3[\sin\delta - \delta][\sin\delta + \delta]}\right). \end{aligned}$$

If the true frequencies $\hat{\omega}_1$ and $\hat{\omega}_2$ are separated by two steps in the discrete Fourier transform, $\delta = 2\pi$, we may reasonably ignore all but the δ^4 term to obtain

$$\begin{aligned} b_{11} &\approx \frac{(\hat{A}_1^2 + \hat{A}_3^2)N^3}{24} \\ b_{22} &\approx \frac{(\hat{A}_2^2 + \hat{A}_4^2)N^3}{24} \\ b_{12} &\approx \frac{(\hat{A}_1\hat{A}_2 + \hat{A}_3\hat{A}_4)N^3\sin(\delta)}{8\delta}. \end{aligned}$$

Having the mixed partial derivatives we may now apply the general formalism (4.14) to obtain

$$(\omega_1)_{\text{est}} = \hat{\omega}_1 \pm \sqrt{\frac{48\sigma^2}{N^3(\hat{A}_1^2 + \hat{A}_3^2)\left\{1 - \frac{9(\hat{A}_1\hat{A}_2 + \hat{A}_3\hat{A}_4)^2\sin^2(\delta)/\delta^2}{4(\hat{A}_1^2 + \hat{A}_3^2)(\hat{A}_2^2 + \hat{A}_4^2)}\right\}}}$$

$$(\omega_2)_{\text{est}} = \hat{\omega}_2 \pm \sqrt{\frac{48\sigma^2}{N^3(\hat{A}_2^2 + \hat{A}_4^2)\left\{1 - \frac{9(\hat{A}_1\hat{A}_2 + \hat{A}_3\hat{A}_4)^2\sin^2(\delta)/\delta^2}{4(\hat{A}_1^2 + \hat{A}_3^2)(\hat{A}_2^2 + \hat{A}_4^2)}\right\}}}$$

The accuracy estimates reduce to (6.18) when the frequencies are well separated. When the frequencies have approximately the same amplitudes and δ is order of 2π (the frequencies are separated by two steps in the fast Fourier transform) the interaction term is down by approximately 1/36; and one expects the estimates to be nearly the same as those for a single frequency. Probability theory indicates that two frequencies which are as close together as two steps in a discrete Fourier transform do not interfere with each other in any significant way. Also note the appearance of the sinc function in the above estimates. When the frequencies are separated by a Nyquist step ($|\omega_1 - \omega_2| = 2\pi/N$) the frequencies cannot interfere with each other. Although this is a little surprising at first sight, a moment's thought will convince one that when the frequencies are separated by $2\pi/N$ the sampled vectors are exactly orthogonal to each other and because we are effectively taking dot products between the model and the data, of course they cannot interfere with each other.

6.3.3 More Accuracy Estimates

To better understand the maximum theoretical accuracy with which two frequencies can be estimated we have prepared Table 6.2. To make these estimates comparable to those obtained in Chapter 2 we have again assumed $N = 1000$ data points and $\sigma = 1$. There are three regions of interest: when the frequency separation is small compared to a single step in the discrete Fourier transform; when the separation is of order one step; and when the separation is large. Additionally we would like to understand the behavior when the signals are of the same amplitude, when one signal is slightly larger than the other, and when one signal is much larger than the other. When we prepared this table we used the joint posterior probability of two frequencies (3.16) assuming the variance σ^2 known. The estimates obtained are the "best" in the sense that in a real data set with $\sigma = 1$, and $N = 1000$ data points the accuracy estimates one obtains will be, nearly always, slightly worse than those contained in table 6.2.

The three values of $(\omega_1 - \omega_2)$ examined correspond to $\delta = 1/4$, $\delta = 4$, and $\delta = 16$: roughly these correspond to frequency separations of 0.07, 0.3, and 5.1 Hz. We held the squared magnitude of one signal constant, and the second is either 1, 4 or 128 times larger.

When the separation frequency is 0.07 Hz the frequencies are indistinguishable. The smaller component cannot be estimated accurately. As the magnitude of the

Table 6.2: Two Frequency Accuracy Estimates

$\frac{\sqrt{B_2^2+B_4^2}}{\sqrt{B_1^2+B_3^2}}$	$\Delta f = 0.07$ Hz		$\Delta f = 0.3$ Hz		$\Delta f = 5.1$ Hz	
	$\delta\hat{f}_1$ Hz	$\delta\hat{f}_2$ Hz	$\delta\hat{f}_1$ Hz	$\delta\hat{f}_2$ Hz	$\delta\hat{f}_1$ Hz	$\delta\hat{f}_2$ Hz
1	±0.091	±0.091	±0.027	±0.027	±0.025	±0.025
4	±0.091	±0.088	±0.027	±0.013	±0.025	±0.012
128	±0.091	±0.034	±0.025	±0.0024	±0.025	±0.0022

We ran a number of simulations to determine how well two frequencies could be determined. In column 1 the two frequencies are separated by only 0.07 Hz and cannot be resolved. In column 2 the separation frequency is now 0.3 Hz and the resolution is approximately 0.0025 Hz for each of the three amplitudes tested. We would have to move one of the frequencies by 11 standard deviations before they would overlap each other. In column 3 the frequencies are separated by 5.1 Hz and we would have to move one of the frequencies by 200 standard deviations before they overlapped.

second signal increases, the estimated accuracy of the second signal becomes better as one's intuition would suppose it should (the signal looks more and more like one frequency). But even at 128:1 probability theory still senses that all is not quite right for a single frequency, and gives an accuracy estimate wider than for a true one frequency signal. However, for very close frequencies the true resolving power is conveyed only by the two-dimensional plot like Fig. 6.10 below; not by the numbers in Table 6.2.

When the separation frequency is 0.3 Hz or about one step in the discrete Fourier transform, the accuracy estimates indicate that the two frequencies are well resolved. By this we mean one of the frequencies would have to be moved by 11 standard deviations before it would be confounded with the other (two parameters are said to be confounded when probability theory cannot distinguish their separate values). This is true for all sample signals in the table; it does, however, improve with increasing amplitude. According to probability theory, when two frequencies are as close together as one Nyquist step in the discrete Fourier transform, those frequencies are clearly resolvable by many standard deviations even at $S/N = 1$; the Rayleigh criterion is far surpassed.

When the separation frequency is 5.1Hz, the accuracy estimates determine both frequencies slightly better. Additionally, the accuracy estimates for the smaller frequency are essentially 0.025Hz which is the same as the estimate for a single harmonic frequency that we found previously (2.12). Examining Table 6.2 more carefully, we see that when the frequencies are separated by even a single step in the discrete Fourier transform, the accuracy estimates are essentially those for the single harmonic frequencies. The ability to estimate two close frequencies accurately is essentially independent of the separation frequency, as long as it is greater than or approximately equal to one step in the discrete Fourier transform!

6.3.4 The Power Spectral Density

The power spectral density (4.16) specifically assumed there were no confounded parameters. The exchange symmetry, in the two-harmonic frequency problem, ensures there are two equally probable maxima. We must generalize (4.16) to account for these. The generalization is straightforward. We have from (4.16)

$$\hat{p}(\{\omega\}) \approx 4\overline{h^2}\frac{P(\{\omega\}|D,\langle\sigma^2\rangle,I)}{\int d\{\omega\}P(\{\omega\}|D,\langle\sigma^2\rangle,I)}. \tag{6.22}$$

The generalization is in the approximating of $P(\{\omega\}|D,\langle\sigma^2\rangle,I)$. Suppose for simplicity that the two frequencies are well separated and the variance σ^2 is known; then the matrix b_{jk} becomes

$$b_{jk} = -\frac{m}{2}\frac{\partial^2\overline{h^2}}{\partial\omega_j^2}\delta_{jk}.$$

Which gives

$$\begin{aligned}\frac{P(\{\omega\}|D,\langle\sigma^2\rangle,I)}{\int d\{\omega\}P(\{\omega\}|D,\langle\sigma^2\rangle,I)} &\approx \left(\frac{b_{11}}{2\pi\langle\sigma^2\rangle}\frac{b_{22}}{2\pi\langle\sigma^2\rangle}\right)^{\frac{1}{2}} \\ &\times \exp\left\{-\frac{b_{11}}{2\langle\sigma^2\rangle}(\hat{\omega}_1-\omega_1)^2-\frac{b_{22}}{2\langle\sigma^2\rangle}(\hat{\omega}_2-\omega_2)^2\right\}\end{aligned}$$

when expanded around $\omega_1 \approx \hat{\omega}_1$, $\omega_2 \approx \hat{\omega}_2$, and

$$\begin{aligned}\frac{P(\{\omega\}|D,\langle\sigma^2\rangle,I)}{\int d\{\omega\}P(\{\omega\}|D,\langle\sigma^2\rangle,I)} &\approx \left(\frac{b_{11}}{2\pi\langle\sigma^2\rangle}\frac{b_{22}}{2\pi\langle\sigma^2\rangle}\right)^{\frac{1}{2}} \\ &\times \exp\left\{-\frac{b_{11}}{2\langle\sigma^2\rangle}(\hat{\omega}_1-\omega_2)^2-\frac{b_{22}}{2\langle\sigma^2\rangle}(\hat{\omega}_2-\omega_1)^2\right\}\end{aligned}$$

when we expand around the other maximum. But to be consistent we must retain the same symmetries in the approximation to the probability density as it originally possessed: the approximation which is valid everywhere is

$$\frac{P(\{\omega\}|D,\langle\sigma^2\rangle,I)}{\int d\{\omega\}P(\{\omega\}|D,\langle\sigma^2\rangle,I)} \approx \frac{1}{2}\left(\frac{b_{11}}{2\pi\langle\sigma^2\rangle}\frac{b_{22}}{2\pi\langle\sigma^2\rangle}\right)^{\frac{1}{2}}$$

$$\times \left[\exp\left\{-\frac{b_{11}}{2\langle\sigma^2\rangle}(\hat{\omega}_1-\omega_1)^2-\frac{b_{22}}{2\langle\sigma^2\rangle}(\hat{\omega}_2-\omega_2)^2\right\}\right.$$

$$\left.+ \exp\left\{-\frac{b_{11}}{2\langle\sigma^2\rangle}(\hat{\omega}_1-\omega_2)^2-\frac{b_{22}}{2\langle\sigma^2\rangle}(\hat{\omega}_2-\omega_1)^2\right\}\right]$$

The factor of 1/2 comes about because there are two equally probable maxima. The power spectral density is a function of both ω_1 and ω_2, but we wish to plot it as a function of only one variable ω. We can do this by integrating out the nuisance parameter (in this case one of the two frequencies). From symmetry, it cannot matter which frequency we choose to eliminate. We choose to integrate out ω_1 and to relabel ω_2 as ω. Performing this integration we obtain

$$\hat{p}(\omega) \approx 2\overline{h^2}(\hat{\omega}_2,\omega)\sqrt{\frac{b_{11}}{2\pi\langle\sigma^2\rangle}}\exp\left\{-\frac{b_{11}}{2\langle\sigma^2\rangle}(\hat{\omega}_1-\omega)^2\right\}$$

$$+ 2\overline{h^2}(\hat{\omega}_1,\omega)\sqrt{\frac{b_{22}}{2\pi\langle\sigma^2\rangle}}\exp\left\{-\frac{b_{22}}{2\langle\sigma^2\rangle}(\hat{\omega}_2-\omega)^2\right\}$$

and using the fact that

$$\overline{h^2}(\hat{\omega}_1,\hat{\omega}_2)=\overline{h^2}(\hat{\omega}_2,\hat{\omega}_1)=C(\hat{\omega}_1)+C(\hat{\omega}_2)$$

we have

$$\hat{p}(\omega) \approx 2[C(\hat{\omega}_1)+C(\hat{\omega}_2)]\left[\sqrt{\frac{b_{11}}{2\pi\langle\sigma^2\rangle}}\exp\left\{-\frac{b_{11}}{2\langle\sigma^2\rangle}(\hat{\omega}_1-\omega)^2\right\}\right.$$

$$\left.+ \sqrt{\frac{b_{22}}{2\pi\langle\sigma^2\rangle}}\exp\left\{-\frac{b_{22}}{2\langle\sigma^2\rangle}(\hat{\omega}_2-\omega)^2\right\}\right].$$

We see now just what that exchange symmetry is doing: The power spectral density conveys information about the total energy carried by the signal, and about the accuracy of each line, but the two terms have equal areas; it contains no information about how much energy is carried in each line. That is not too surprising; after all we defined the power spectral density as the total energy carried by the signal per unit $\{\omega\}$. That is typically what one is interested in for an arbitrary model function.

However, the multiple frequency problem is unique in that one is typically interested in the power carried by each line; not the total power carried by the signal. This is not really a well defined problem in the sense that as two lines become closer and closer together the frequencies are no longer orthogonal and power is shared between them. The problem becomes even worse when one considers nonstationary frequencies. We will later define a line spectral density which will give information about the power carried by one line when there are multiple well separated lines in the spectrum.

6.3.5 Example – Two Harmonic Frequencies

To illustrate some of the points we have been making about the two-frequency probability density (6.17) we prepared a simple example, Fig. 6.9. This example was prepared from the following equation

$$d_i = \cos(0.3i + 1) + \cos(0.307i + 2) + e_i$$

where e_i has variance one and the index runs over the symmetric time interval ($-255.5 \leq i \leq 255.5$) by unit steps. This time series, Fig. 6.9(A), has two simple harmonic frequencies separated by approximately 0.6 steps in the discrete Fourier transform. One step corresponds to $|\hat{\omega}_1 - \hat{\omega}_2| \approx 2\pi/512 = 0.012$.

From looking at the raw time series one might just barely guess that there is more going on than a simple harmonic frequency plus noise, because the oscillation amplitude seems to vary slightly. If we were to guess that there are two close frequencies, then by examining the data one can guess that the difference between these two frequencies is not more than one cycle over the entire time interval. If the frequencies were separated by more than this we would expect to see beats in the data. If there are two frequencies, the second frequency must be within 0.012 of the first (in dimensionless units). This is in the region where the frequency estimates are almost but not quite confounded.

Now Fig. 6.9(B) the periodogram (continuous curve) and the fast Fourier transform (open circles) show only a single peak. The single frequency model has estimated a frequency which is essentially the average of the two. Yet the two frequency posterior probability density Fig. 6.10 shows two well resolved, symmetrical maxima. Thus the inclusion of just this one simple additional fact – that the signal may have two frequencies – has greatly enhanced our ability to detect the two signals. Prior information, even when it is only qualitative, can have a major effect on the quantitative

Figure 6.9: Two Harmonic Frequencies – The Data

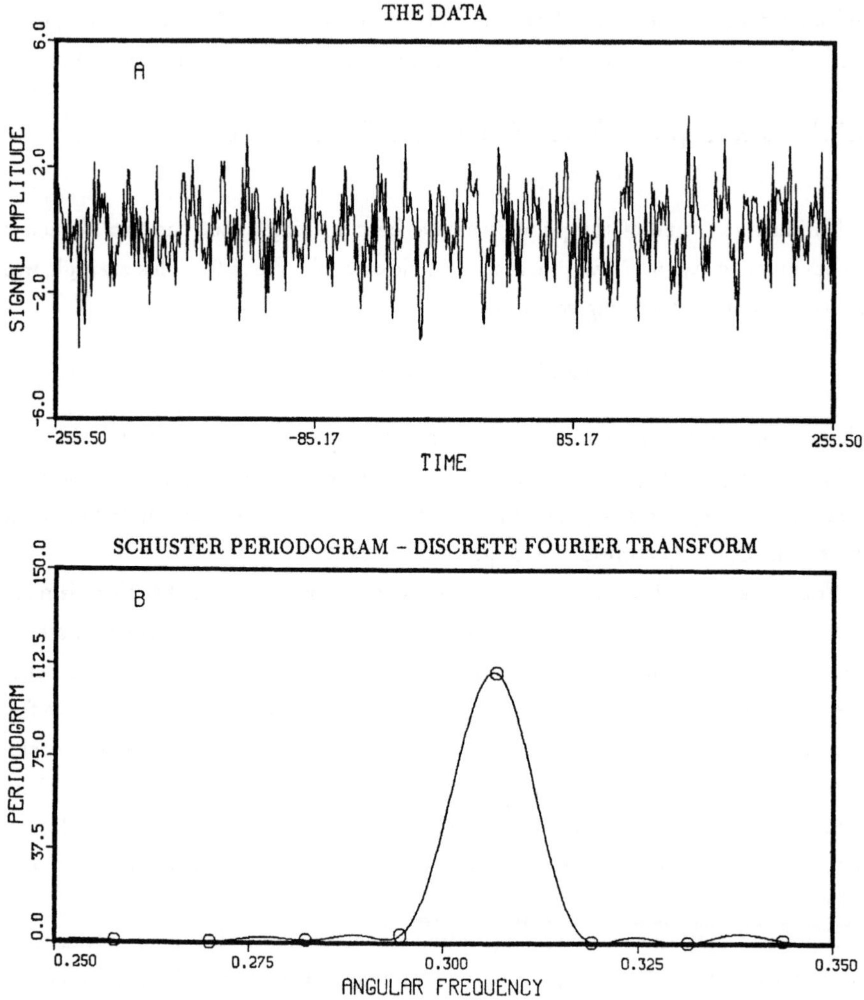

The data (A) contain two frequencies. They are separated from each other by approximately a single step in the discrete Fourier transform. The periodogram (B) shows only a single peak located between the two frequencies.

Figure 6.10: Posterior Probability density of Two Harmonic Frequencies

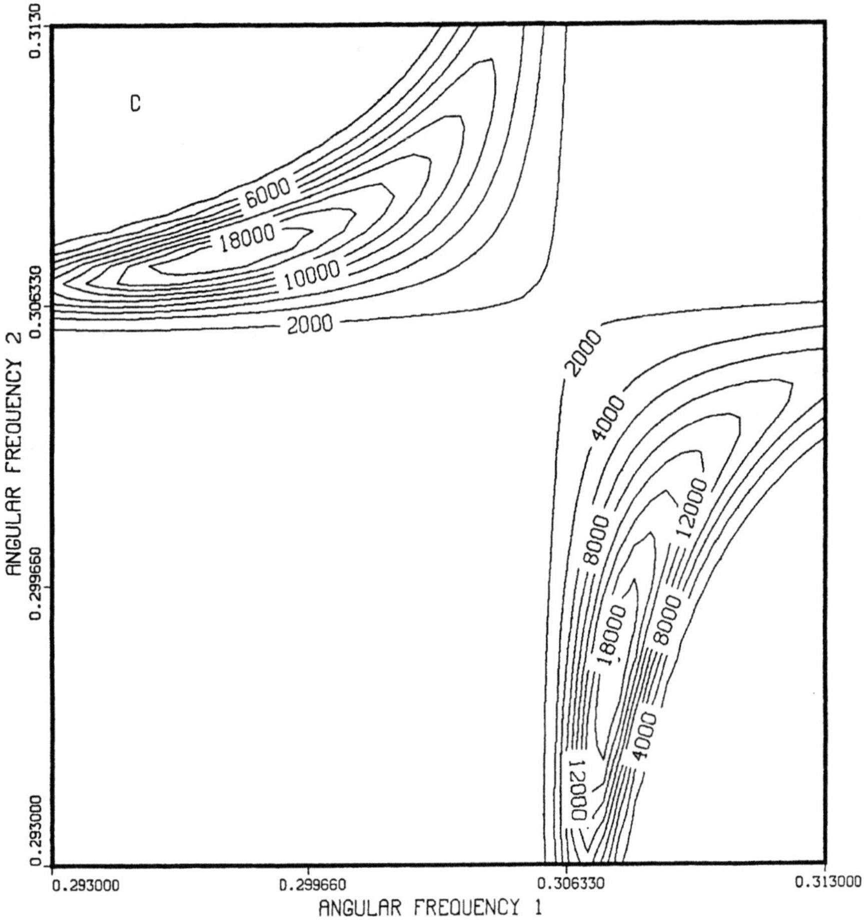

This is a fully normalized posterior probability density of two harmonic frequencies in the data, Fig. 6.9. The two-frequency probability density clearly indicates the presence of two frequencies. The posterior odds ratio prefers the two-frequency model by 10^7 to 1.

conclusions we are able to draw from a given data set.

This plot illustrates numerically some of the points we have been making. First, in the two harmonic frequency probability density there are three discrete Fourier transforms: one along each axis, and a third along $\omega_1 = \omega_2$. The two transforms along the axes form ridges. If the frequencies are very close and have the same amplitude the ridges are located at the average of the two frequencies: $0.5(0.3 + 0.307) = 0.335$. The discrete Fourier transform along the line of symmetry $\omega_1 = \omega_2$ can almost be imagined. As we approach the true frequencies, $\omega_1 \approx 0.307$ and $\omega_2 \approx 0.3$, these ridges have a slight bend away from the value indicated by the discrete Fourier transform: these very close frequencies are not orthogonal. When the true frequencies are well separated, these ridges intersect at right angles (the cross derivatives are zero) and the frequencies do not interfere with each other. Even now, two very close frequencies do not interfere greatly.

According to probability theory, the odds in favor of the two-frequency model compared to the one-frequency model are 10^7 to 1. Now that we know the data contain two partially resolved frequencies, we could proceed to obtain data over a longer time span and resolve the frequencies still better. Regardless, it is now clear that what one can detect clearly depends on what question one asks, and thus on what prior information we have to suggest the best questions.

6.4 Estimation of Multiple Stationary Frequencies

The problem of estimating multiple stationary harmonic frequencies can now be addressed. The answer to this problem is, of course, given by the "Student t-distribution" Eq. (3.17) using

$$f(t) = \sum_{j=1}^{r} B_j \cos \omega_j t + \sum_{j=1}^{r} B_{r+j} \sin \omega_j t \tag{6.23}$$

as a model. No exact analytic solution to this problem exists for more than a few frequencies. However, a number of interesting things can be learned by studying this problem.

6.5 The "Student t-Distribution"

We begin this process by calculating the g_{ij} matrix explicitly. For a uniformly sampled time series this is given by

$$g_{jk} = \begin{pmatrix} c_{11} & c_{12} & \cdots & c_{1r} & 0 & \cdots & \cdots & 0 \\ c_{21} & c_{22} & \cdots & c_{2r} & \vdots & \vdots & \vdots & \vdots \\ \vdots & \vdots & \ddots & \vdots & \vdots & \vdots & \vdots & \vdots \\ c_{r1} & c_{r2} & \cdots & c_{rr} & 0 & \cdots & \cdots & 0 \\ 0 & \cdots & \cdots & 0 & s_{11} & s_{12} & \cdots & s_{1r} \\ \vdots & \vdots & \vdots & \vdots & s_{21} & s_{22} & \cdots & s_{2r} \\ \vdots & \vdots & \vdots & \vdots & \vdots & \vdots & \ddots & \vdots \\ 0 & \cdots & \cdots & 0 & s_{r1} & s_{r2} & \cdots & s_{rr} \end{pmatrix}$$

where c_{jk} and s_{jk} were defined earlier (6.11, 6.12). To investigate the full solution we first make the same large N approximations we made in the two-frequency problem.

When the frequencies are well separated, $|\omega_j - \omega_k| \gg 2\pi/N$, the diagonal elements are again replaced by $N/2$ and the off diagonal elements are given by $B(\omega_j, \omega_k)/2$, using the notation $B_{jk} \equiv B(\omega_j, \omega_k)$ defined earlier by Eq. (6.20). This simplifies the g_{jk} matrix somewhat:

$$g_{jk} = \frac{1}{2} \begin{pmatrix} N & B_{12} & \cdots & B_{1r} & 0 & \cdots & \cdots & 0 \\ B_{21} & N & \cdots & B_{2r} & \vdots & \vdots & \vdots & \vdots \\ \vdots & \vdots & \ddots & \vdots & \vdots & \vdots & \vdots & \vdots \\ B_{r1} & B_{r2} & \cdots & N & 0 & \cdots & \cdots & 0 \\ 0 & \cdots & \cdots & 0 & N & B_{12} & \cdots & B_{1r} \\ \vdots & \vdots & \vdots & \vdots & B_{21} & N & \cdots & B_{2r} \\ \vdots & \vdots & \vdots & \vdots & \vdots & \vdots & \ddots & \vdots \\ 0 & \cdots & \cdots & 0 & B_{r1} & B_{r2} & \cdots & N \end{pmatrix}$$

The problem separates into finding the eigenvalues and eigenvectors of an $r \times r$ matrix.

Multiple Well-Separated Frequencies

For convenience, assume the frequencies are ordered: $\omega_1 < \omega_2 < \cdots < \omega_r$. The exchange symmetries in this problem ensure that we can always do this. Now assume that $|\omega_j - \omega_k| \gg 2\pi/N$. The g_{jk} matrix will simplify significantly, because all of the off-diagonal elements are essentially zero:

$$g_{jk} \approx \frac{N}{2}\delta_{jk}.$$

The problem has separated completely, and the joint posterior probability of multiple harmonic frequencies when the variance is known is given by

$$P(\omega_1, \cdots, \omega_r | D\sigma, I) \propto \exp\left\{\sum_{j=1}^{r} \frac{C(\omega_j)}{\sigma^2}\right\}.$$

This result was first found by Jaynes [12]. The estimated frequencies are the r largest peaks in the discrete Fourier transform, again in agreement with common sense. Of course, the accuracy estimates of the frequencies are those obtained from the single harmonic frequency problem.

If one were to estimate the accuracy from multiple frequency data using a single-frequency model, the answers would not be the same; the estimated noise variance σ^2 would be far greater, because multiple-frequency data will not fit a single-frequency model. Thus in realistic cases where the noise variance σ^2 must be estimated from the data, it is essential to use the estimated variance from the multiple-frequency model even when the frequencies are well separated.

The results from this section let us see the discrete Fourier transform in yet another way: the discrete Fourier transform is a sufficient statistic for the estimation of multiple-well separated harmonic frequencies. P. Whittle [29] derived the periodogram from the principle of maximum likelihood in 1954 and stated that "... in practice the periodogram presents a wildly irregular appearance, suggesting little or nothing to the eye." It now appears that this depends on the condition of the brain behind that eye; after a little Bayesian education, a periodogram suggests a great deal to the eye because one knows where to look. When the frequencies are well separated, it is only the very largest peaks in the periodogram that are important for frequency estimation. The common practice of taking the log of the periodogram is just about the worst thing one could do, because it accents the noise and suppresses information about the frequencies.

Two Close Frequencies

Now assume that the first two frequencies are close together: $|\omega_1 - \omega_2| \approx 2\pi/N$. Then the off diagonal term B_{12} is not small. But by assumption all the remaining off-diagonal terms are negligible. The problem separates into a two-frequency problem for the close frequencies and $r-1$ one-frequency problems. A feasible procedure for estimating multiple harmonic frequencies is now clear. We calculate the probability of a single harmonic frequency in the data. We take the single largest peak from the data and we examine it with a two-frequency model. If there is any evidence of two frequencies, we will obtain a better fit; if not the frequencies will confound with each other. Now generate the best model function from the estimated parameters (either a one or two-frequency model) and subtract it from the data. The difference is the residual signal which must be analyzed further.

What we are contemplating here is, in spirit, what an economist would call detrending (i.e. estimating a trend and then subtracting it from the data). Normally this is a bad thing to do, because the trend and the signal of interest are not orthogonal. We can do this here because the orthogonality properties of multiple harmonic model functions ensures that the error is small. But, we stress, it is only the special properties of the sine and cosine functions that make this possible.

Next we examine the residual signal using the same procedure. We compute the posterior probability of a frequency in the residuals and examine the largest peak for two frequencies. We repeat the entire procedure until we have reduced the residuals to noise (i.e. until they exhibit no visible regularity). Determining the stopping place is not generally a problem, but if there are many small signals present it will be necessary to use the procedures developed in Chapter 5 to determine the total number of frequencies present. We stress again that this procedure is only applicable to the multiple stationary frequency problem and then only because of the special properties of the sine and cosine functions. Even here, if there is evidence of multiple close frequencies it will be necessary to use the estimates obtained from this procedure as initial estimates for a full multiple-frequency analysis on the data.

6.5.1 Example – Multiple Stationary Frequencies

To illustrate some of the points we have been making we have prepared a simple example of a stationary signal with multiple harmonic frequencies. This simple example was prepared from

$$\begin{aligned} f(t) &= \cos(0.1i+1) + 2\cos(0.15i+2) + 5\cos(0.3i+3) \\ &+ 2\cos(0.31i+4) + 3\cos(1i+5) + e_i \end{aligned}$$

and shown in Fig. 6.11(A), where e_i has unit variance and there are $N = 512$ data points. The periodogram Fig. 6.11(B) resolves the four well-separated frequencies and then hints that the frequency near 0.3 could be two frequencies. To estimate these frequencies we simply postulated a five-frequency model and used the estimates from the periodogram as initial estimates of the frequencies. We located the maximum of the five-dimensional posterior probability density and determined the accuracy estimate using the procedure given in (4.14).

The estimated frequencies and amplitudes from these data are

frequency	amplitude
0.0998±0.0001	0.9±0.08
0.1498±0.0002	2.08±0.08
0.3001±0.0002	4.97±0.08
0.3102±0.0001	1.95±0.08
0.9999±0.0001	2.92±0.08

These are in excellent agreement with the true values. The estimated noise variance for these data is 0.98 and the true variance is 1.0. For this data set with $N = 512$ data values, the "best" estimate for the well-separated frequencies is given by (2.10)

$$\omega_{\text{est}} \approx \hat{\omega} \pm \sqrt{48\sigma^2/N^3(B_1^2 + B_2^2)}$$

which gives

frequency	amplitude
0.1000±0.0006	1.0±0.08
0.1500±0.0002	2.0±0.08
0.3000±0.0001	5.0±0.08
0.3100±0.0002	2.0±0.08
1.0000±0.0002	3.0±0.08

and the actual values we obtained are all comparable to these: in some cases a little better and others a little worse.

6.5.2 The Power Spectral Density

We saw in the two-frequency problem that the power spectral density $\hat{p}(\omega)$ is telling us something about the energy density of the signal and about the accuracy of the line. We have not generalized that function to account for the symmetry properties of the multiple frequency problem, and we do that now. The generalization is straightforward and we simply give the result for well-separated frequencies here.

When the frequencies are well separated the problem essentially splits into a series of one-frequency problems: all we must do is to maintain the symmetries of the original probability density. Maintaining those symmetries and integrate out all but

Figure 6.11: Multiple Harmonic Frequencies

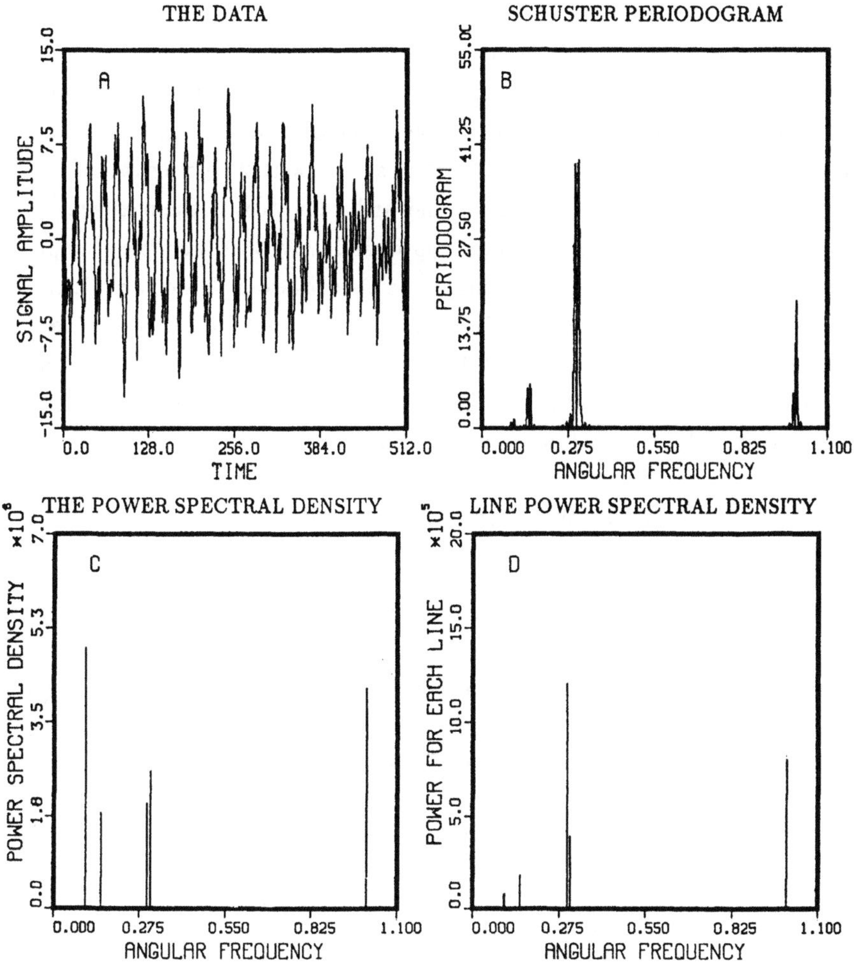

The data (A) contain five frequencies. Three of the five are well separated. The Schuster periodogram (B) resolves the three well-separated frequencies, but one cannot tell if the peak near $\omega = 0.3$ is one or two frequencies. The power spectral density $\hat{p}(\omega)$ (C) clearly separates all five frequencies while the height is indicative of the resolution. The height of the line power spectral density (D) is indicative of the energy carried by the line.

one frequency, the power spectral density may be approximated as

$$\hat{p}(\omega) \approx 2\left[\sigma^2 + \sum_{j=1}^{r} C(\hat{\omega}_j)\right] \sum_{k=1}^{r} \left[\frac{b_{kk}}{2\pi\sigma^2}\right]^{\frac{1}{2}} \exp\left\{-\frac{b_{kk}(\hat{\omega}_k - \omega)^2}{2\sigma^2}\right\}.$$

As was stressed earlier this function expresses information about the total energy carried by the signal and about the accuracy of each line, but nothing about the power carried by one line. After the fact this is not too surprising: after all, we asked a question about the total energy carried by the signal, and not a question about the power carried by one line. If we wish information about the power carried by one line, we must ask a question about one line, and we do that in the next subsection. First we illustrate the generalized power spectral density with a simple example. In addition to determining the frequencies in the previous example we have plotted the power spectral density in Fig. 6.11(C). We see from (C) that the five frequencies have been well resolved by the "Student t-distribution": the widths of the lines from (C) are indications of how well the lines have been determined from the data while the integral over all lines is the total energy carried by the signal in the observation time.

6.5.3 The Line Power Spectral Density

We would like to plot a power spectral density that is an indication of the power carried by the individual spectral lines. This is easily done simply by defining the appropriate spectral density. Here we define a line power spectral density $\hat{S}(\omega)$ as the posterior expected value of the energy carried by one sinusoidal component of the signal in the frequency range $d\omega$. This is given by

$$\begin{aligned} \hat{S}(\omega) &= \frac{N}{2}\int (B_1^2 + B_{1+r}^2) P(\{B\}, \{\omega\}|\sigma, D, I) dB_1 \cdots dB_m d\omega_2 \cdots d\omega_r \\ &= \frac{2}{r}\left[\sigma^2 + C(\omega)\right] \sum_{k=1}^{r} \left[\frac{b_{kk}}{2\pi\sigma^2}\right]^{\frac{1}{2}} \exp\left\{-\frac{b_{kk}(\hat{\omega}_k - \omega)^2}{2\sigma^2}\right\} \end{aligned}$$

where we have performed the integrals over all but the first frequency. We have relabeled ω_1 as ω. When we computed this expectation value we used the amplitudes for frequency ω_1 however, the exchange symmetries in this problem ensure we will obtain the same result whichever one we chose to leave behind. This is essentially just the marginal posterior probability density normalized to the power carried by one spectral line. The integral over ω will give the total energy carried by all of the spectral lines, and in this approximation each line contributes its total energy to the integral. We have included an example of this in the multiple frequency example given earlier see Fig. 6.11(D). In this figure the lines are normalized to a power level, the heights are indications of the total energy carried by a line, and the width is an indication of the accuracy of the estimates.

6.6 Multiple Nonstationary Frequency Estimation

The problem of multiple nonstationary frequencies is easily addressed using the "Student t-distribution" (3.17). As with the multiple stationary frequency estimation problem, an analytic solution is not feasible for more than a few frequencies. However, we already know that this problem separates. If it did not, the discrete Fourier transform would not be useful on this problem; and it is.

The way to handle this problem is to apply the "Student t-distribution" numerically. One can apply the single-frequency-plus-decay model when the nonstationary frequencies are well separated, and then use more complex models where needed. The numerical procedure to use is to calculate the discrete Fourier transform of the data, and from it compute the logarithm of the probability of a single harmonic frequency. Then set up a nonstationary frequency model using the single best frequency from the discrete Fourier transform. Locate the maximum of the probability density and then compute the residuals. These residuals are essentially what probability theory is calling the noise. Repeat the Fourier transform step on the residuals. If there are additional frequencies in the data, repeat the process using two, three, $\cdots$, frequencies model until all frequencies have been accounted for. Of course, one can save time here by starting with an initial model that has the same number of well-separated peaks as are in the Fourier transform of the data. But care must be taken; if these signals are decaying, one must supply reasonable estimates for the decay rates and this can be very difficult.

When applying this procedure, there is no need to check to see if any of the peaks have multiple frequencies. Later passes through the procedure will resolve double structure. If any of the peaks has multiple frequencies, then when one fits the main peak not all of the signal will be removed, and on some later cycle through the procedure the second frequency will be the largest remaining effect in the data and the procedure will pick it out. The procedure works so well and the effects are so striking, that an example is needed. We give this example in the next chapter.

Chapter 7

APPLICATIONS

Perhaps the greatest test of any theory is not so much how it was derived, but how it works. Here we apply the theory as developed in the preceding chapters to a number of specific examples including: NMR signals, economic time series, and Wolf's relative sunspot numbers. Also, we examine how multiple measurements affect the analysis.

7.1 NMR Time Series

NMR provides an excellent example of how the introduction of modern computers has revolutionized a branch of science. With the aid of computers more data can be taken and summarized into a useful form faster than has ever been possible before. The standard way to analyze an NMR experiment is to obtain a quadrature data set, with two separate measurements, $90°$ out of phase with each other, and to do a complex Fourier transform on these data [30]. The global phase of the discrete complex transform is adjusted until the real part (called an absorption spectrum) is as symmetric as possible. The frequencies and decay rates are then estimated from the absorption spectrum. There are, of course, good physical reasons why the absorption spectrum of the "true signal" is important to physicists. However, as we have emphasized repeatedly since Chapter 2, the discrete Fourier transform is an optimal frequency estimator only when a single simple harmonic frequency is present, and there are no conditions known to the author under which an absorption spectrum will give optimal frequency estimates.

We will apply the procedures developed in the previous sections to a time series from a real NMR experiment, and contrast our analysis to the one done using the absorption spectrum. The NMR data used are of a free-induction decay [31], Fig. 7.1. The sample contained a mixture of 63% liquid Hydrogen-Deuterium (HD) and Deuterium (D_2) at 20.2°K. The sample was excited with a 55MHz pulse, and its response was observed using a standard mixer-modulation technique. The resulting signal is in the audio range where it has several oscillations at about 100Hz. The data were sampled at $\Delta t = 0.0005$ seconds, and $N = 2048$ data points were taken for each channel. The data therefore span a time interval of about one second. As was discussed earlier, we are using dimensionless units. The relation to physical units is given by

$$f = \frac{\omega}{2\pi\Delta t}\text{Hz}, \qquad \text{Period} = \frac{2\pi\Delta t}{\omega}\text{Seconds}$$

where f is the frequency in Hertz, ω is the frequency in radians per step, and Δt is the sampling time interval in seconds.

In these data there are a number of effects which we would like to investigate. First, the indirect J coupling [32] in the HD produces a doublet with a splitting of about 43Hz. The D_2 in the sample is also excited; its resonance is approximately in the middle of the HD doublet. One of the things we would like to determine is the shift of the D_2 singlet relative to the center of the HD doublet. In addition to the three frequencies there are two different characteristic decay times; the decay rate of the HD doublet is grossly different from that of D_2 [32]. However, an inhomogeneous magnetic field could mask the true decay: the decay could be magnet limited. We would like to know how strongly the inhomogeneous magnetic field has affected the decay.

The analysis we did in Chapter 3, although general, did not use a notation appropriate to two channels. We need to generalize the notation; there are two different measurements of this signal, (assumed to be independent), and we designate them as $d_1(t_i)$ and $d_2(t_i)$. The model functions will be abbreviated as $f_1(t)$ and $f_2(t)$ with the understanding that each measurement of the signal has different amplitudes and noise variance, but the same $\{\omega\}$ parameters.

We can write the likelihood (3.2) immediately to obtain

$$L(f_1, f_2) \propto (\sigma_1\sigma_2)^{-N} \exp\left\{-\frac{X}{2\sigma_1^2} - \frac{Y}{2\sigma_2^2}\right\}$$

where

$$X \equiv \sum_{i=1}^{N}[d_1(t_i) - f_1(t_i)]^2$$

Figure 7.1: Analyzing NMR Spectra

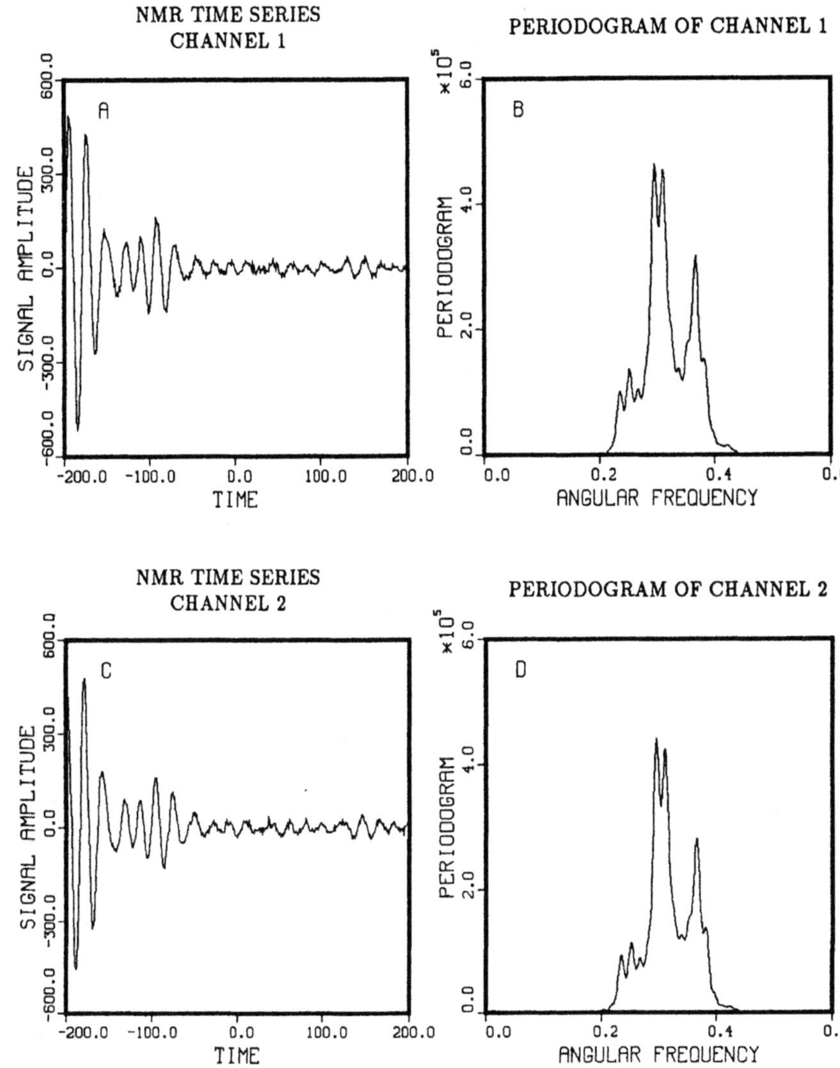

The data are channel 1 (A) and 2 (C) from a quadrature detected NMR experiment. The time series or free-induction decay is of a sample containing a mixture of D_2 and HD in a liquid phase. Theory indicates there should be three frequencies in these data: A D_2 singlet, and an HD doublet with a 43Hz separation. The singlet should be approximately in the center of the doublet. In the discrete Fourier transform, (B channel 1) and (D channel 2), the singlet appears to be split.

$$Y \equiv \sum_{i=1}^{N}[d_2(t_i) - f_2(t_i)]^2.$$

Because the amplitudes and noise variance are assumed different in each channel, we may remove these using the same procedure developed in Chapter 3.

As in most of our examples, this procedure is conservative; if we had definite prior information linking the amplitude or variances in the two channels, we could exploit that information at this point to get still better estimates of the $\{\omega\}$ parameters. For example, if we knew that the noise was strongly correlated in the two channels, that would enable us to estimate the noise in each channel more accurately. After removing the nuisance parameters, the marginal posterior probability of the $\{\omega\}$ parameters is just the product of the "Student t-distributions" Eq. (3.17) for each channel separately:

$$P(\{\omega\}|D,I) \propto \left[1 - \frac{m\overline{h^2}_1}{N\overline{d^2}_1}\right]^{\frac{m-N}{2}} \left[1 - \frac{m\overline{h^2}_2}{N\overline{d^2}_2}\right]^{\frac{m-N}{2}} \tag{7.1}$$

where the subscripts refer to the channel number. As explained previously, (7.1) in effect estimates the noise level independently in the two channels. This procedure is general and can be applied whenever two measurements of a signal are available; it is not restricted to NMR data. It is possible to specialize the estimation procedures to include this quadrature model, as well as the aforementioned phase and noise correlations. If all of this prior information is incorporated into the analysis (the author has, in fact, done this), we would expect to improve the results considerably. However, the present results will prove adequate for most purposes.

A procedure for dealing with the multiple frequency problem was outlined in Chapter VI, and we will apply that procedure here. The first step in any frequency estimation problem is to plot the data and the log of the probability of a single harmonic frequency. If there is only one data channel, this is essentially the periodogram of the data, Fig. 7.1(B) and Fig. 7.1(D). When more than one channel is present, the log probability of a single harmonic frequency is essentially the sum of the periodograms for each channel, weighted by the appropriate reciprocal variances. If the variances are unknown, then the appropriate statistic is the log of (7.1), shown in Fig. 7.2.

Now as was shown in Chapter 6, if the frequencies are well separated, a peak in the periodogram above the noise level is evidence – but not proof – of a frequency near that peak. From examining Fig. 7.2 we see there are nine resolved peaks in $0.2 < \omega < 0.4$ and suggestions of five more unresolved ones. This is many more peaks than theoretical physics indicates there should be. Is this evidence of more

Figure 7.2: The Log_{10} Probability of One Frequency in Both Channels

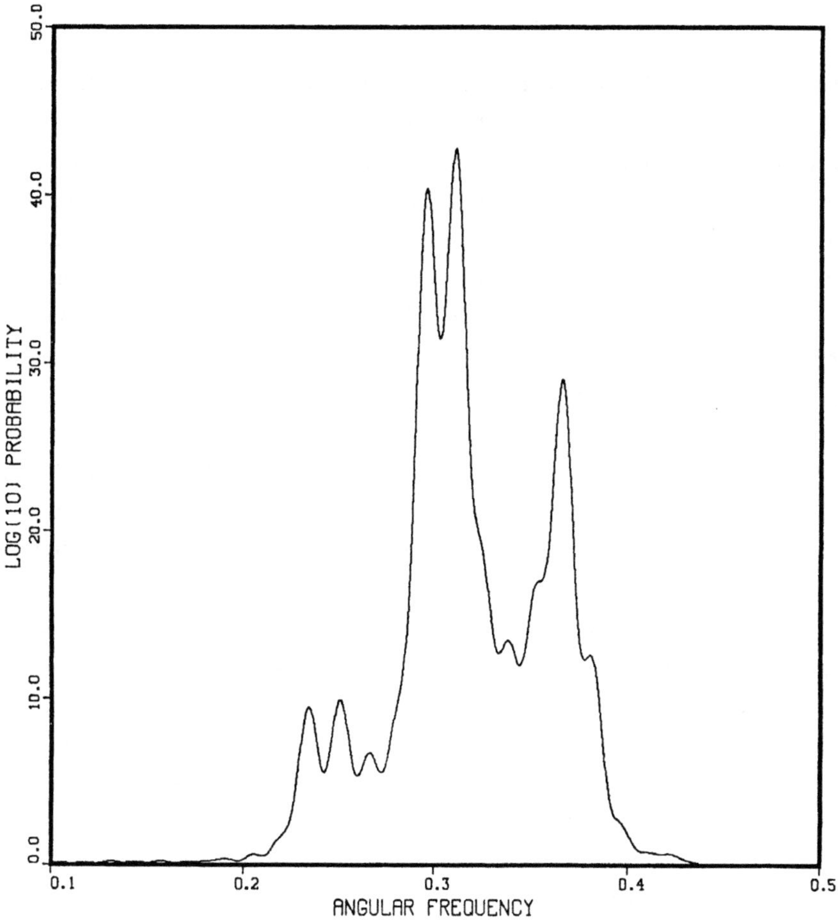

When more than one channel is present, the periodogram is not the proper statistic to be analyzed for indications of a simple harmonic frequency. The proper statistic (shown above) is log of the probability of a single harmonic frequency in both channels.

going on than theory predicts? To answer this question we will apply the general procedure outlined in the preceding chapter for determining multiple frequencies. We first fit the data with the single best frequency plus decay. We choose Lorentzian decay instead of Gaussian because physical theory indicates the decay is Lorentzian in a liquid phase. The model we used is

$$f_1(t) = [B_1 \cos(\omega_1 t) + B_2 \sin(\omega_1 t)]e^{-\alpha t}.$$

The computer code in Appendix E was used to evaluate the "Student t-distribution" Eq. (3.17) for each channel, and these were multiplied to obtain, Eq. (7.1). We searched in the two dimensional parameter space until we located the maximum of the distribution by the "pattern" searching procedure noted before. Next we computed the signal having the predicted parameters. The model (dotted line) and the data from the real channel are shown in, Fig. 7.3(A). It is clear from examining this figure as well as from examining the residuals in, Fig. 7.3(B), that there is at least a second frequency in this data. We see from the probability of a single harmonic frequency in the residuals, Fig. 7.3(C), that there is still strong evidence for additional frequencies near 0.3.

We then proceeded to a two-frequency-plus-decay model and repeated this procedure. That is, we estimated the second frequency plus decay from the residuals, and then used the results from the one-frequency model plus the estimates from the residuals as the initial estimates in a two-frequency model of the original data. We searched this four-parameter space until we located the maximum of the probability density. The results from the two-frequency model are displayed in Fig. 7.4. The model (dotted line) now takes on more characteristics of the signal (A), while the residuals (B) and the probability of a single harmonic frequency in the residuals (C) continue to show evidence for additional effects in the data. Notice the structure of the probability of a single frequency in the residuals. The addition of a second frequency removed one peak and essentially left the others unchanged. We demonstrated in Chapter 6 that when the frequencies are well separated the multiple-frequency estimation problem separates into a series of single-frequency problems, and this just confirms numerically that result.

To compare the two-frequency model to the one-frequency we computed the posterior odds ratio, this is given by:

$$\text{posterior odds} = \frac{P(f_2|I)}{P(f_1|I)} \frac{P(D|f_2, I)}{P(D|f_1, I)}$$

Figure 7.3: The One-Frequency Model

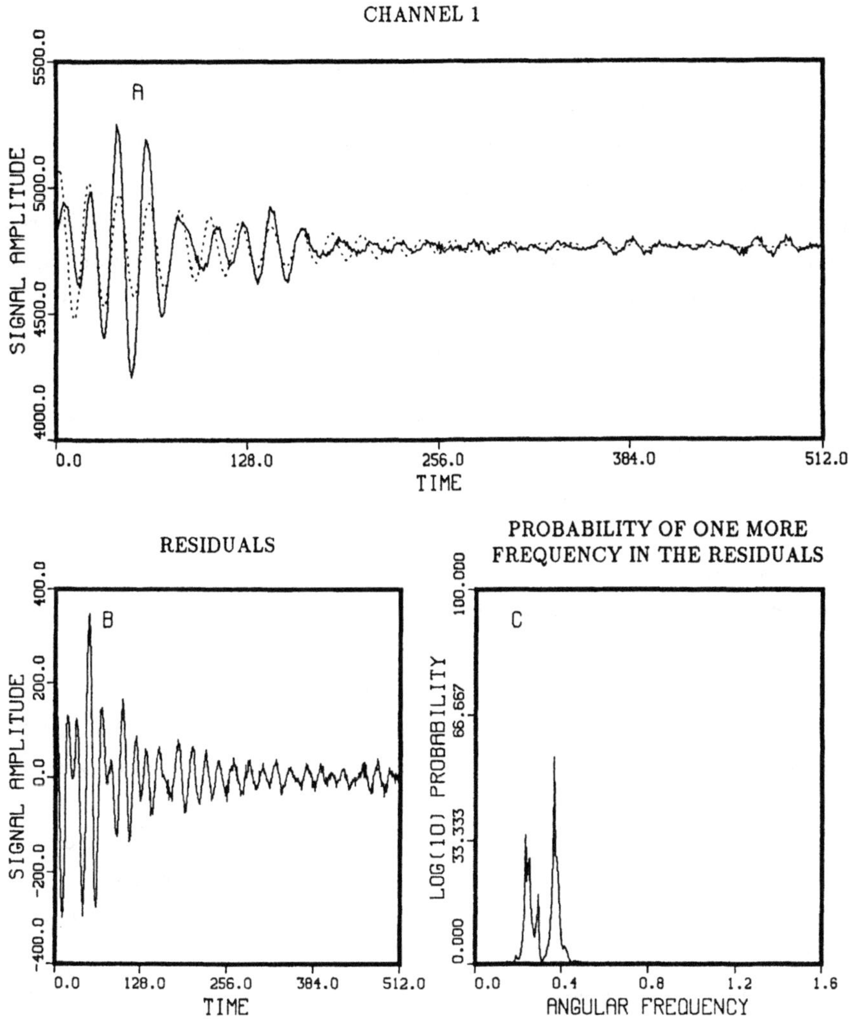

The data from one channel of the NMR experiment (solid line), and a one-frequency-plus-decay model with the predicted parameters (dotted line) are shown in (A). Next we computed the residuals: the differences between the data and the model (B). The residuals clearly indicate additional effects in the data. Last we computed the probability of a single harmonic frequency in the residuals (C). This clearly indicates there are additional effects in the data.

where, $P(f_1|I)$ and $P(f_2|I)$ are the prior probability of the one-frequency and two-frequency models, and $P(D|f_1, I)$ and $P(D|f_2, I)$ are the global likelihoods, Eq. 5.9, for the one-frequency and two-frequency models. We have some prior information about how many frequencies should be present: theoretical physics indicates there should be three frequencies, however, we will assume either of the models is equally probable and set $P(f_2|I)/P(f_1|I) = 1$. We then computed the likelihood ratio and find there is one chance in 10^{457} that the one-frequency model is a better description of the phenomenon than the two-frequency model. There is zero chance that the one-frequency model represents the data better! But it might be that very unusual noise is confusing us, and there is a tiny chance that the one-frequency model would represent the next data set better. To see how tiny that chance is, note that the number of microseconds in the estimated age of the universe is only about 10^{24}.

Then we proceeded to the three-frequency-plus-decay model. The most probable frequency is the low frequency peak in the vicinity of 0.3; so we ran the three-frequency-plus-decay model using this low frequency as the initial estimate for the third frequency, Fig. 7.5. The model has now taken on most of the dominant characteristics of the signal as in Fig. 7.5(A): indeed the triple is the largest effect in the data. However, fitting the triple does not account for the long time behavior of the system. Notice in the residuals, Fig. 7.5(B), that there is still more than enough signal left for the eye to make out the oscillations easily. We see from the probability of a single frequency in the residuals, Fig. 7.5(C), that there is still evidence for additional frequencies in the data. The posterior odds ratio for the two-frequency-plus-decay model compared to the three-frequency-plus-decay model indicates that there is one chance in 10^{703} that the two-frequency model is a better description than the three-frequency model.

To see if there are additional effects in the data we proceeded to the four-frequency-plus-decay model. Figure 7.6(A) is a plot of the data and the model. Now the model is making a much better showing in the long time behavior of the system, but even here we have not accounted for all the effects in the data. Clearly in the residuals, Fig. 7.6(B), there is a small unaccounted for signal; the probability of a single frequency in the residuals, Fig. 7.6(C), verifies this, indicating it to be a high-frequency component (not shown in Fig. 7.6). The posterior odds ratio of the four-frequency-plus-decay model to the three-frequency-plus-decay model indicates there is one chance in 10^{80} that the three-frequency model is a better description than the four-frequency model.

Figure 7.4: The Two-Frequency Model

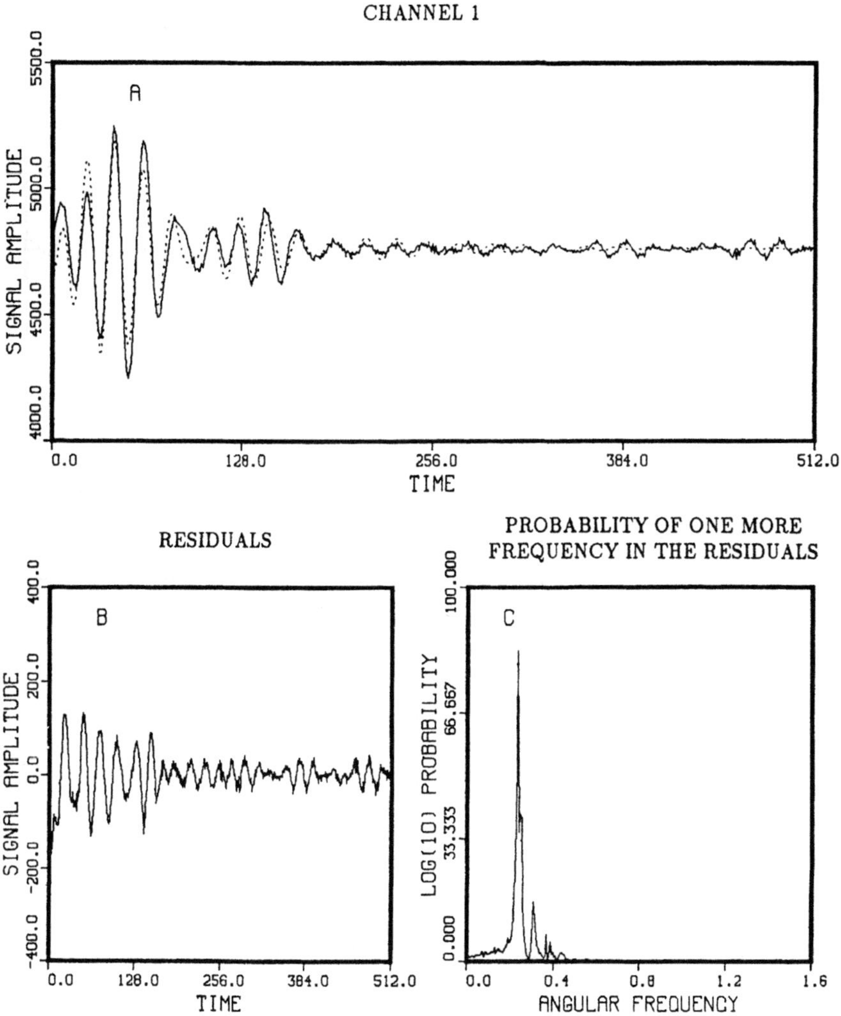

Next we computed the probability of two frequencies plus decay in both channels. The model (dotted line) and the data (solid line) are displayed in (A). The residuals (B) clearly indicate additional effects in the data. We then computed the probability of a single frequency in the residuals and displayed that in panel (C).

Figure 7.5: The Three-Frequency Model

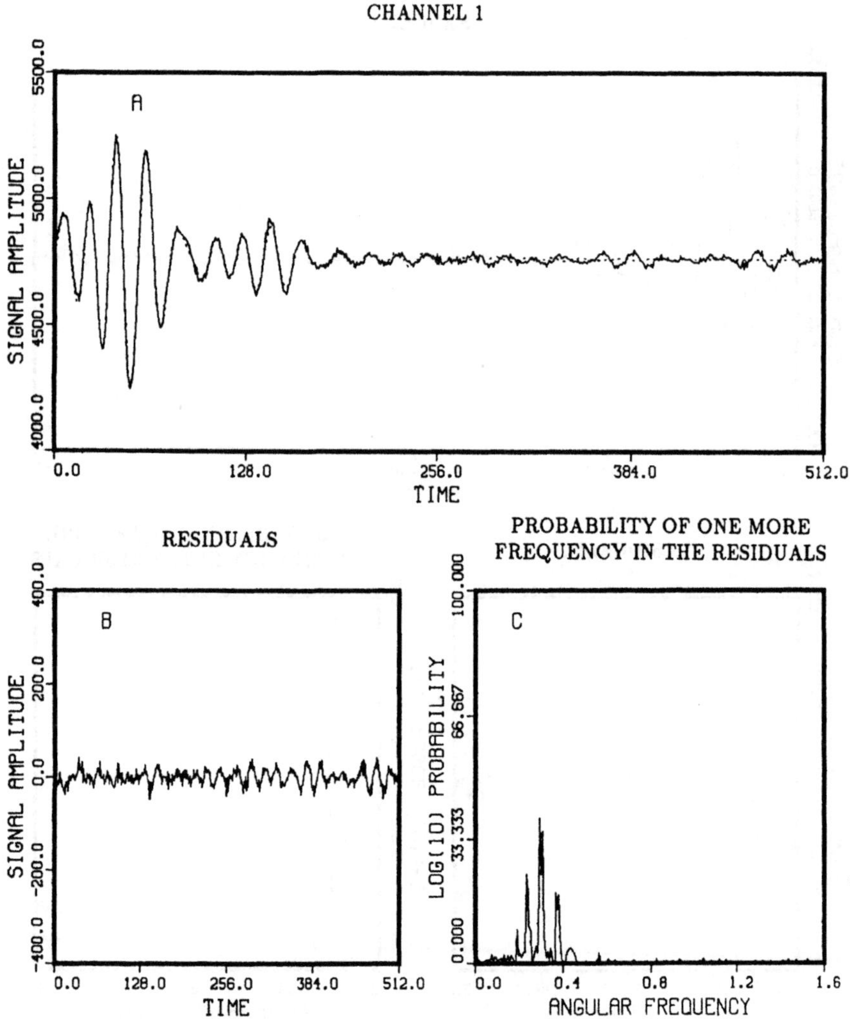

Because the two-frequency-plus-decay model did not take up all of the signal, we proceeded to a three-frequency-plus-decay model, (A) dotted line. The residuals (B) clearly indicate additional effects in the data. We computed the probability of a single frequency in the residuals and displayed that in panel (C). Again we see there are additional effects in this data set.

Figure 7.6: The Four-Frequency Model

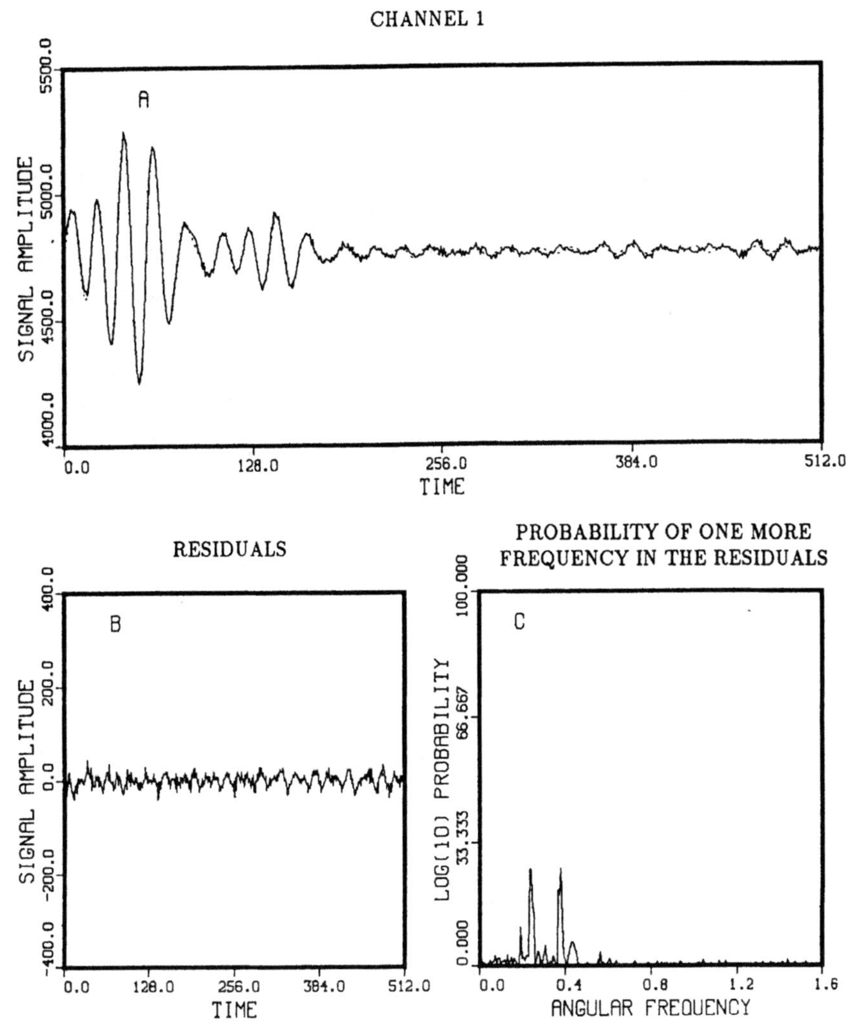

Because the three-frequency-plus-decay model did not account for all of the signal, we proceeded to a four-frequency-plus-decay model, (A) dotted line. The residuals (B) continue to indicate additional effects in the data. We computed the probability of a single frequency in the residuals and displayed that in panel (C). Again we see that there are additional effects in this data set.

We continued repeating this procedure until we accounted for all systematic components – see Fig. 7.7 through Fig. 7.9. Now in Fig. 7.9(C) the residuals are finally beginning to look like Gaussian white noise. However, there is some evidence for a very small additional frequency, see Fig. 7.9(C). We did not go further because this frequency, although present in the real channel, is not present to any significant degree in the quadrature data channel.

Our probability analysis indicates there are at least seven frequencies in these data, of which one is attributable to the instrumentation. That leaves six frequencies located near 0.3 in dimensionless units. The posterior probability of a single harmonic frequency in the combined data, Fig. 7.2, gives evidence of multiple complex phenomena around 0.3 but it could not sort out what is going on. This is not too surprising, given that there are six frequencies in this region. The one-frequency model has done surprisingly well. The absorption spectrum, Fig. 7.10(A), on the other hand, shows only three peaks in this region. This simple example illustrated that the discrete Fourier transform gives evidence of frequencies in the data that an absorption spectrum does not. Although the probability of a single harmonic frequency or the Schuster periodogram is not an exact estimator for multiple frequencies, it is adequate as long as the frequencies are well separated. The only time one must worry about this statistic being incorrect is when the frequencies are close together (as they were here). But by contrast, there are no conditions under which the absorption spectrum is an optimal frequency estimator, and the global phase adjustment on the absorption spectrum can suppress indications of frequencies in the data.

We developed the procedures for estimating the accuracy of the frequencies and the amplitudes and we have used those procedures here [to apply them we calculated the second derivatives numerically (4.13)]. The results of this calculation are:

Frequency Hertz	Decay Rate Hertz	Amplitude Real	Amplitude Imaginary
75.0695 ± 0.0005	7.294 ± 0.003	49	46
78.1231 ± 0.0002	19.613 ± 0.001	170	160
94.1207 ± 0.0008	8.569 ± 0.001	71	72
98.0187 ± 0.0001	23.211 ± 0.001	354	318
117.6052 ± 0.0001	16.336 ± 0.001	193	188
121.0824 ± 0.0002	11.270 ± 0.001	67	66

We also estimated the signal-to-noise ratio, Eq. (4.8), for each channel:

$$\frac{\text{Signal}}{\text{Noise}} = 1606 \text{ in channel 1,}$$

Figure 7.7: The Five-Frequency Model

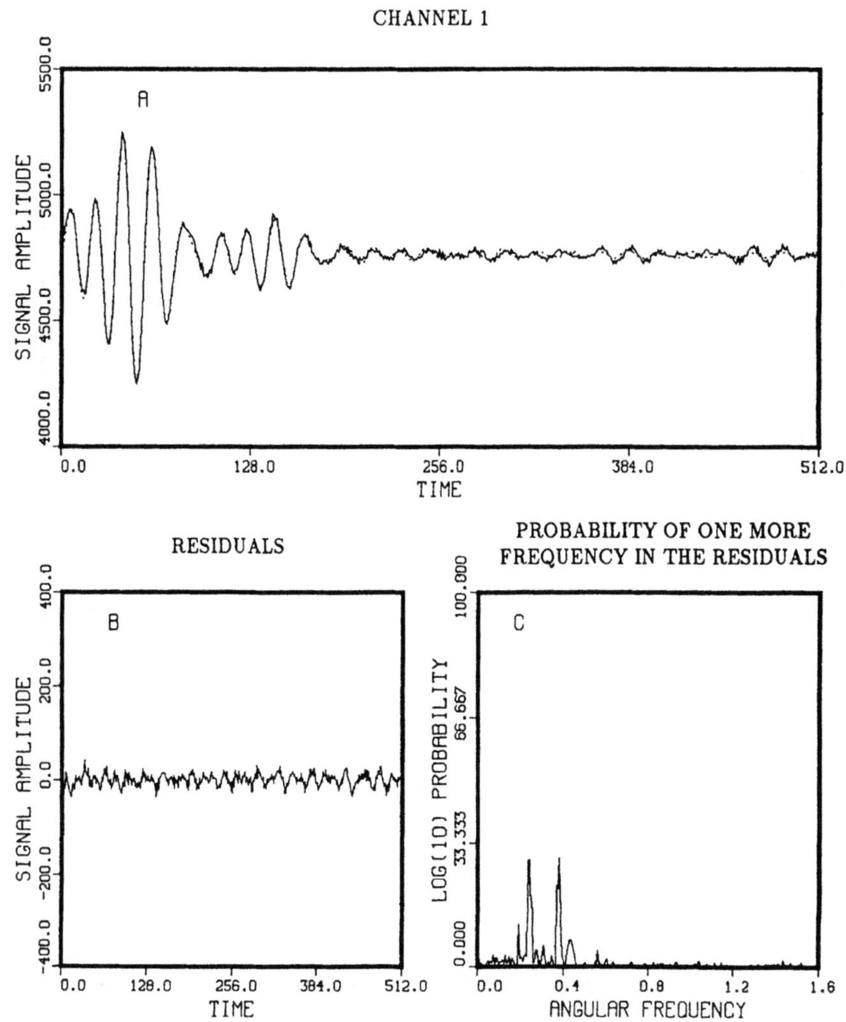

Because the four-frequency model did not account for all of the signal, we proceeded to a five-frequency model, (A) dotted line. The residuals (B) continue to indicate additional effects in the data. We computed the probability of a single frequency in the residuals and displayed that in panel (C). There is no apparent change in (C) because a very high frequency component was remove by the four-frequency model. Again we see that there are additional effects in this data set.

Figure 7.8: The Six-Frequency Model

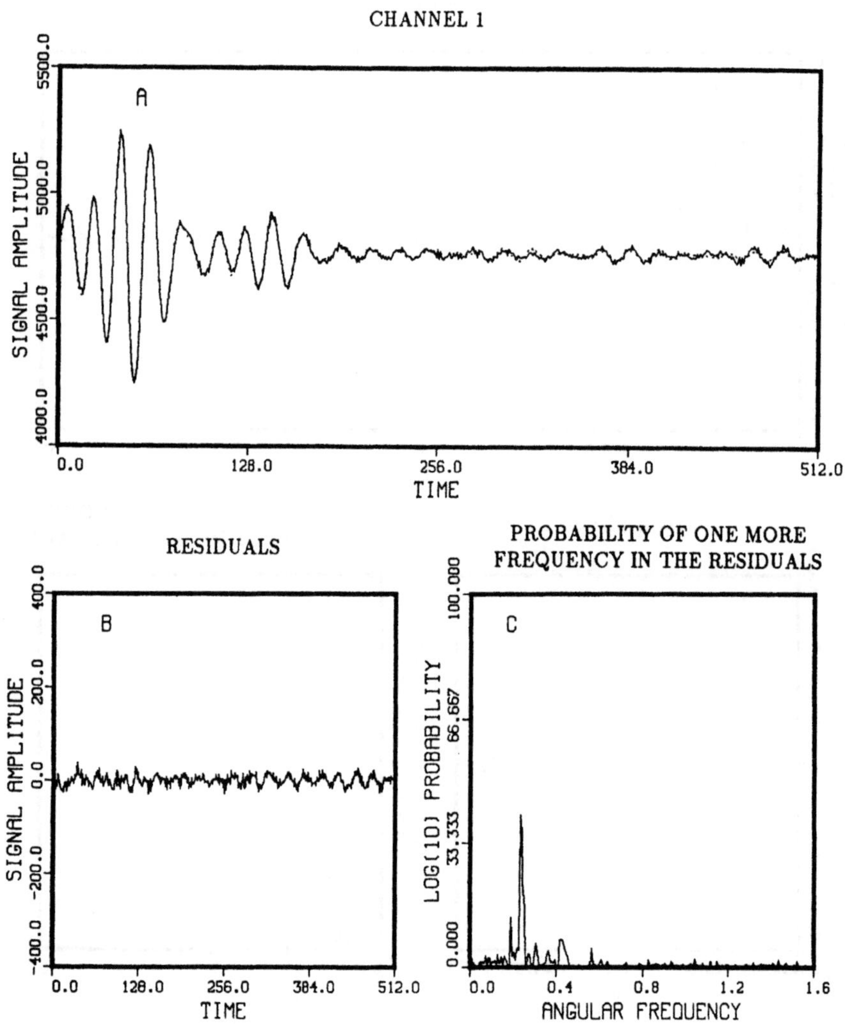

Because the six-frequency model did not account for all of the signal, we proceeded to a seven-frequency model, (A) dotted line. The residuals (B) clearly indicate additional effects in the data. We computed the probability of a single frequency in the residuals and displayed that in panel (C). Again we see there are additional effects in this data set.

Figure 7.9: The Seven-Frequency Model

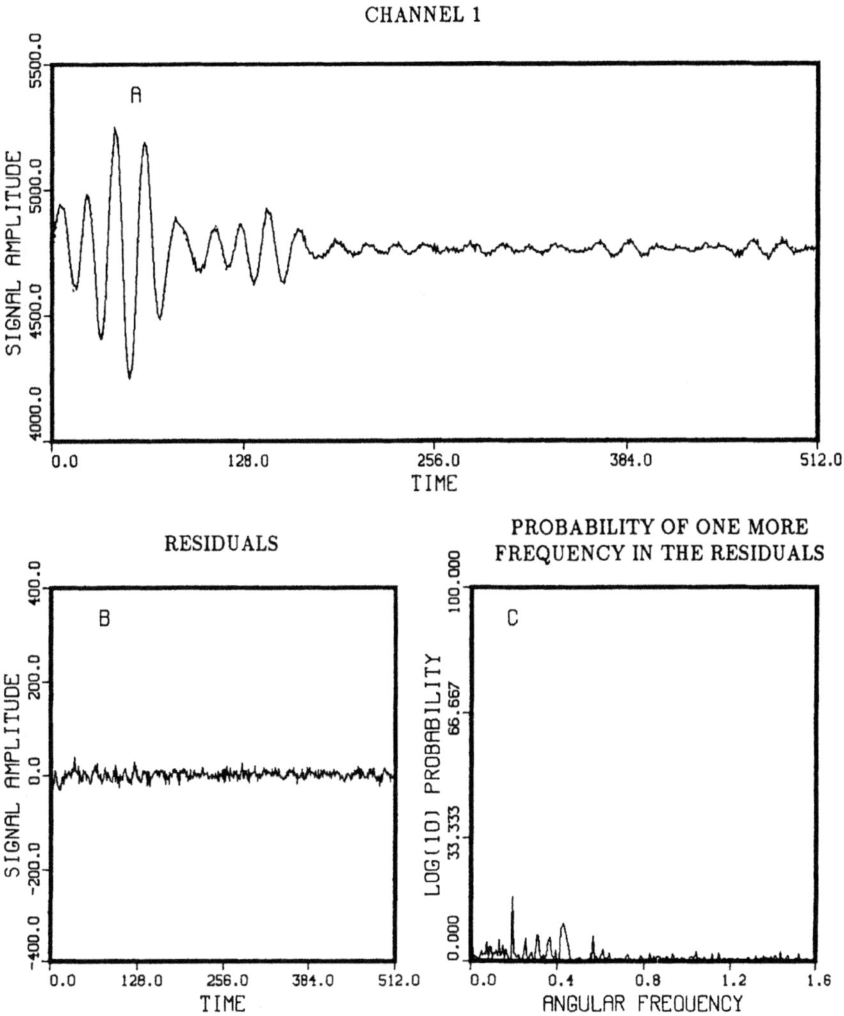

At last, with the seven-frequency model we reached a point where the model and the signal look essentially identical (A). The residuals (B), now look much more like white noise. We computed the probability of a single frequency in the residuals and displayed that in panel (C). Again we see there are additional very small effects in this data set. However, these effects are not repeated in both channels: we interpret these effects be an instrumental artifact.

Figure 7.10: Comparison to an Absorption Spectrum

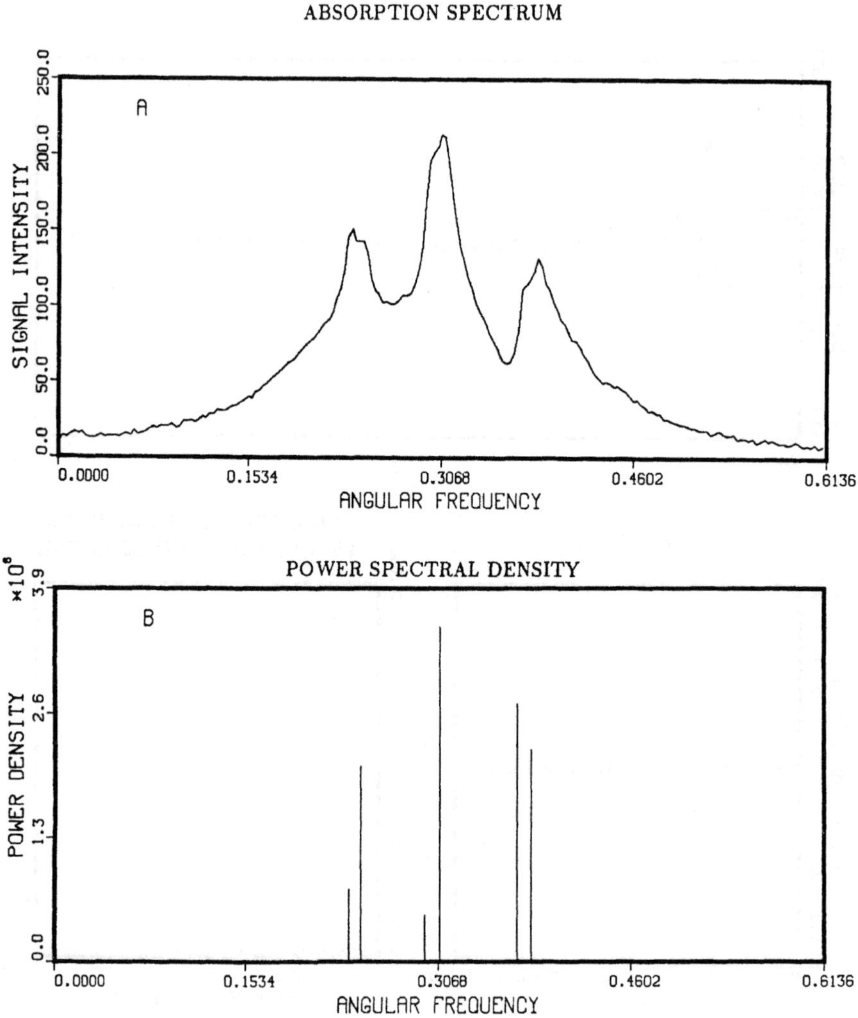

The absorption spectrum (described in the text, see page 117) gives a clear indication of three frequencies and hints at three others (A). Using the full width at half maximum of the absorption spectrum to determine the accuracy estimate and converting to physical units, it determines the frequencies to within ±15Hz. The probability analysis (B) used a seven-frequency model with decay. The estimated accuracy is approximately ±0.001Hz.

$$\frac{\text{Signal}}{\text{Noise}} = 1478 \text{ in channel 2,}$$

and the estimated standard deviation (4.6):

$$(\sigma)_{\text{est}} = 9 \text{ in channel 1,}$$

$$(\sigma)_{\text{est}} = 9 \text{ in channel 2.}$$

The amplitudes were estimated separately in each channel, and if the spectrometer is working correctly we expect the amplitude of each sinusoid to be approximately the same. This serves as an additional check on the model; if we were fitting an appreciable amount of noise, the estimated amplitudes would be different in the two channels.

The quantities of interest are the splitting between the two components of the HD doublet as well as the shift in the center frequency. But physical theory indicated there should be only three frequencies in the region of the main resonance: we find six. The calculation indicates there is clearly more going on here than physical theory indicates there should be. One of the major assumptions made in NMR is that the magnetic field is uniform over the sample. If it is not, the resonances will be spread out, corresponding to different intensities of the local field, and false structure may appear. Here we may be seeing this effect. However, the sharpness of the peaks suggests that the effect is real, conceivably arising from impurities in the sample or from association effects (such as H_4O_2 molecules) not considered in the theory.

However, we have derived a model of the process as if there were two major regions in the sample where the field was approximately uniform. If we wish to derive the splittings we must use the frequencies corresponding to uniform field. In each of the regions where the field is a uniform, the frequency shifts should be according to theory. Thus for the set of frequencies shifted to lower values (75, 94, and 117) the HD doublet separation is

$$\text{High - Low} = 42.536 \pm 0.001\text{Hz}$$

and the center frequency (94 Hz) is displaced from the center of the doublet by 2.217±0.001 Hz. For the set of frequencies (78, 98, and 121) shifted to higher values we have

$$\text{High - Low} = 42.956 \pm 0.001\text{Hz}$$

and the center frequency (98 Hz) is displaced from the center of the doublet by 1.521±0.001 Hz. Both of these tentative answers are in good agreement with the simple theory; unfortunately, until the field shimming problems are cleared up we do not know which to believe, if either. The center frequency is displaced from the center of the doublet in the correct direction, and in reasonable agreement with prior measurements of this quantity [33]. In order to answer these questions it would be necessary to rerun the experiment with better shimming. Additionally, the estimates could be improved somewhat by sampling the data faster.

If one attempts to analyze these data using the standard absorption spectrum Fig. 7.10(A) only three peaks are found, with hints of three other frequencies. The splitting of the HD doublet is approximately correct, but the center peak is shifted in the wrong direction. We can compare these estimates directly to the absorption spectrum. The reason the analysis of this experiment is so difficult with the absorption spectrum is that the full-width at half maximum for the D_2 peak, Fig. 7.10(A), is 15Hz. But this width is indicative only of the decay rates; not the accuracy with which the oscillations frequency is determined. Probability theory has enabled us to separate these entirely different quantities. Figure 7.10(B) gives the estimates from Eq. (7.1). We have plotted these estimates as normalized Gaussians, each centered at the estimated frequency and having the same standard deviation as the estimated frequency. Clearly, the resolution of these frequencies is much improved compared to an absorption spectrum or a discrete Fourier transform. With separately normalized distributions, the heights in Fig. 7.10(B) are indications of the accuracy of the estimates, not of the power carried by the signal.

The accuracy of this procedure may be a little disturbing. To understand it, look at the estimated signal-to-noise ratio in these data. It is on the order of 1500 for each channel. There is essentially nothing in these data sets that can be ignored. Every little bump and wiggle in the discrete Fourier transform is indicative of some effect in the data, and must be accounted for. Because the accuracy of the estimates is inversely proportional to the signal-to-noise of the data, the estimates are very precise. It is rather the inaccuracy of the conventional method that should be disturbing to one.

7.2 Corn Crop Yields

Economic data are hard to analyze, in part because the data are frequently contaminated by large spurious effects, which one does not know how to capture in a model, and the time series are often very short. Here we will examine one example of economic data to demonstrate how to remove some unknown and spurious effects. In particular, we will analyze one hundred year's worth of corn crop data from three states (Kansas, South Dakota, and Nebraska), Fig. 7.11(A) through Fig. 7.11(C) [34]. We would like to know if there is any indication of periodic behavior in these data.

These data have been analyzed before. Currie [35] used a high pass filter and then applied the Burg algorithm [36] to the filtered data. Currie finds one frequency near 20 years which is attributed to the lunar 18.6 year cycle, and another at 11 years, which is attributed to the solar cycle.

There are three steps in Currie's analysis that are troublesome. First, the Burg algorithm is not optimal in the presence of noise (although it is for the problem it was formulated to solve). The fact that it continues to work means that the procedure is reasonably robust; that does not change the fact that it is fundamentally not appropriate to this problem [36]. Second, one has doubts about the filter: could it suppress the effect one is looking for or introduce other spurious effects? Third, to apply the Burg algorithm when the data consist of the actual values of a time series, the autoregression order (maximum lag to be used) must be chosen and there is no theoretical principle to determine this choice. We do not mean to imply that Currie's result is incorrect; only that it is provisional. We would like to apply probability theory as developed in this work to check these results.

The first step in a harmonic analysis is simply to plot the data, Fig. 7.11(A) through Fig. 7.11(C) and the log of the posterior probability of a single harmonic frequency. In the previous example we generalized the analysis for two channels. The generalization to an arbitrary number of channels is just a repeat of the previous arguments:

$$P(\{\omega\}|D,I) \propto \prod_{j=1}^{r} \left[1 - \frac{m_j\overline{h^2}_j}{N_j\overline{d^2}_j}\right]^{\frac{m_j-N_j}{2}} \tag{7.2}$$

where the subscripts refer to the jth channels: each of the models has m_j amplitudes, and each data set contains N_j data values. Additionally it was assumed that the noise variance σ_j was unknown and possibly different for each channel. The "Student t-distributions" Eq. (3.17) for each channel should be computed separately, thus estimating and eliminating the nuisance parameters particular to that channel, and then multiplied to obtain the posterior probability for the common effects, Eq. (7.2). Again if we had prior knowledge of correlations in the "noise" for different channels, we could exploit that information to get better final results, at the cost of more

Figure 7.11: Corn Crop Yields for Three Selected States

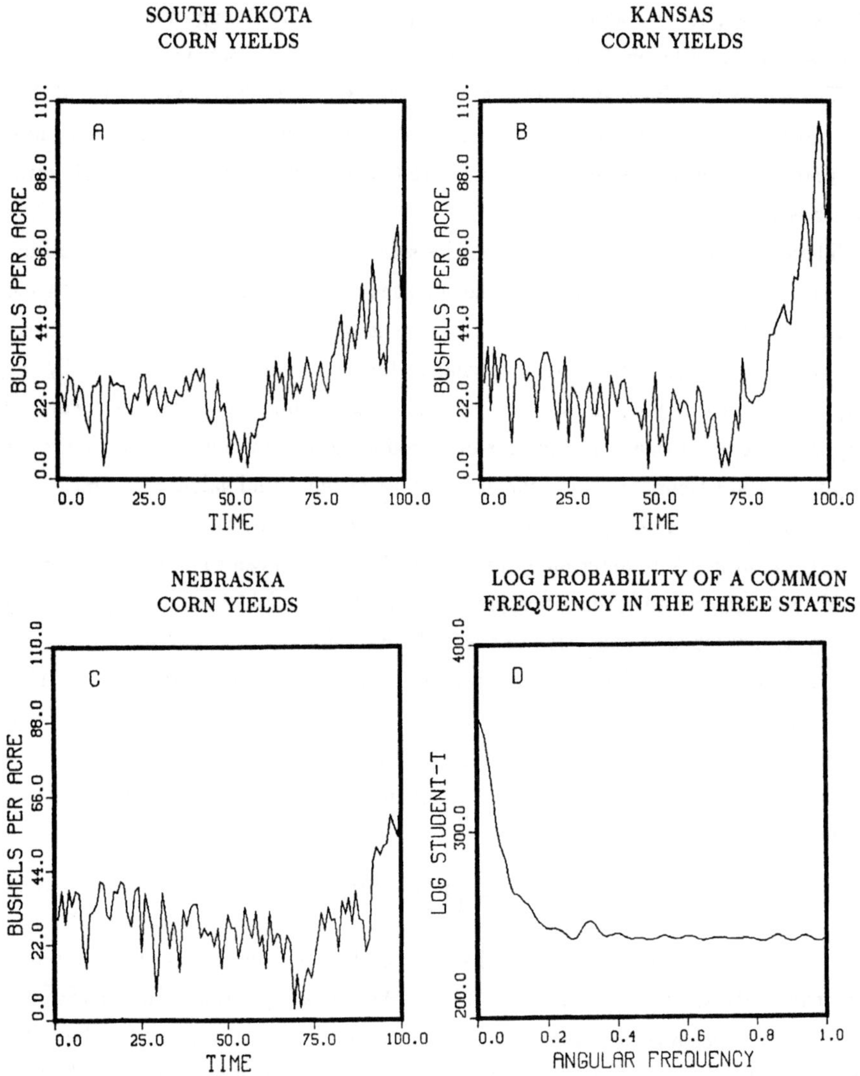

The three data sets analyzed were corn yields in bushels per acre for South Dakota (A), Kansas (B), and Nebraska (C). The log probability of a single common frequency plus a constant is plotted in (D). The question we would like to answer is "Is that small bump located at approximately 0.3, corresponding to a 20 year period, a real indication of a frequency or is it an artifact of the trend?"

computation.

For this harmonic analysis we take the model to be a single sinusoid which oscillates about a constant. The model for the jth channel may be written

$$f_j(t) = B_{j,1} + B_{j,2}\sin(\omega t) + B_{j,3}\cos(\omega t). \tag{7.3}$$

Here we have three channels, named "South Dakota", "Kansas", and "Nebraska". We allow $B_{j,1}$, $B_{j,2}$, and $B_{j,3}$ to be different for each channel; thus there are a total of nine amplitudes, one frequency, and three noise variances. To compute the posterior probability for each measurement, we used the computer code in Appendix E. The log of each "Student t-distribution" Eq. (3.17) was computed and added to obtain the log of the posterior probability of a single common harmonic frequency, Fig. 7.11(D).

What we would like to know is, "Are those small bumps in Fig. 7.11(D) indications of periodic behavior, or are they artifacts of the noise or trend?" To attempt to answer this, consider the following model function

$$f_j(t) = T_j(t) + B_{j,1}\cos(\omega t) + B_{j,2}\sin(\omega t)$$

where we have augmented the standard frequency model by a trend $T_j(t)$. The only parameter of interest is the frequency ω. The trend $T_j(t)$ is a nuisance function; to eliminate it we expand the trend in orthonormal polynomials $L_j(t)$. These orthonormal polynomials could be any complete set. We use the Legendre polynomials with an appropriate scaling of the independent variable to make them orthonormal on the region $(-49.5 \leq t \leq 49.5)$. This is the range of values used for the time index in the sine and cosine terms. After expanding the trend, the model function for the jth measurement can be written

$$f_j(t) = \sum_{k=0}^{E} B_{j,k+1}L_k(t) + B_{j,E+2}\cos(\omega t) + B_{j,E+3}\sin(\omega t).$$

Notice that if the expansion order E is zero the problem is reduced to the previous problem (7.3).

The expansion order E must be set to some appropriate value. From looking at these data one sees that it will take at least a second order expansion to remove the trend. The actual expansion order for the trend is unknown. However, it will turn out that the estimated frequencies are insensitive to the expansion order, as long as the expansion is sufficient to represent the trend without representing the signal of interest. Of course, different orders could have very different implications about other questions than the ones we are asking; for example, predicting the future trend. That is an altogether more difficult problem than the one we are solving.

The effects of increasing the expansion order E can be demonstrated by plotting the posterior probability for several expansion orders - see Fig. 7.12(A)–7.12(H). For expansion order zero, Fig. 7.12(A), through expansion order 2, Fig. 7.12(C) the trend has not been removed: the posterior probability continues to pick out the low frequency trend. When a third order trend is used, Fig. 7.12(D), a sudden change in the behavior is seen. The frequency near $\omega \approx 0.31$ suddenly shows up, along with a spurious low-frequency effect due to the trend. In expansion orders four through seven, Fig. 7.12(E) through Fig. 7.12(H), the trend has been effectively removed and the posterior probability indicates there is a frequency near 0.31 corresponding to a 20.4 year period.

The amount of variability in the frequency estimates as a function of the expansion order will show how strongly the trend expansion is affecting the estimated frequency. The frequency estimates for the fourth through seventh order expansions are

$$\begin{aligned}
(f_4)_{\text{est}} &= 20.60 \quad \pm 0.16 \text{ years} \\
(f_5)_{\text{est}} &= 20.47 \quad \pm 0.18 \text{ years} \\
(f_6)_{\text{est}} &= 20.20 \quad \pm 0.14 \text{ years} \\
(f_7)_{\text{est}} &= 20.47 \quad \pm 0.18 \text{ years.}
\end{aligned}$$

Here the estimated errors represent two standard deviations. Thus, given the spread in the estimates it appears there is indeed evidence for a frequency of a period 20.4 $\pm$ 0.2 years.

Now that the effects of removing a trend are better understood, we can proceed to a two-frequency model plus a trend to see if we can verify Currie's two frequency results. Figure 7.13 is a plot of the log of this probability distribution after removing a fifth order trend. The behavior of this plot is the type one would expect when a two-frequency model is applied to a data set that contains only one frequency. From this we cannot verify Currie's results. That is, for the three states taken as a whole these data show evidence for an oscillation near 20.4 years as he reports, but we do not find evidence for an 11 year cycle. This does not say that Currie's result is incorrect; he incorporated much more data into his calculation, and to check it we would need to include data from at least a dozen more states. While this is a worthy project, it is beyond the scope of this simple demonstration, whose main purpose is to show the good performance of the "theoretically correct" method of trend removal.

Figure 7.12: The Joint Probability of a Frequency Plus a Trend

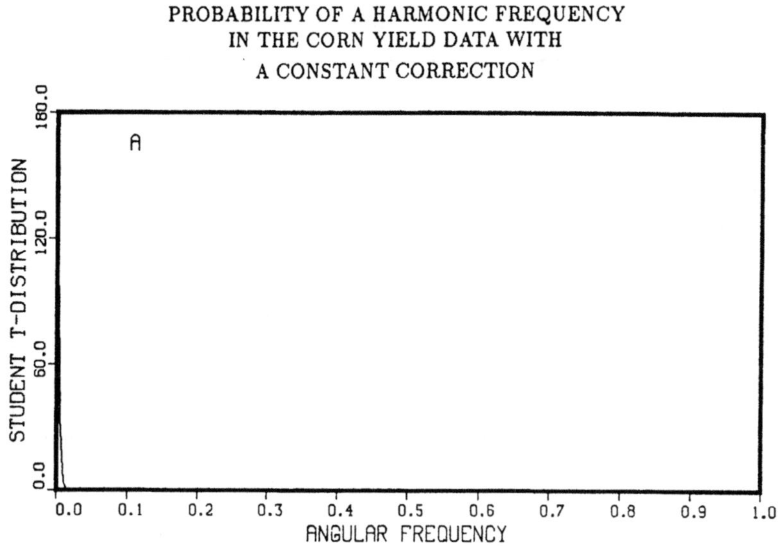

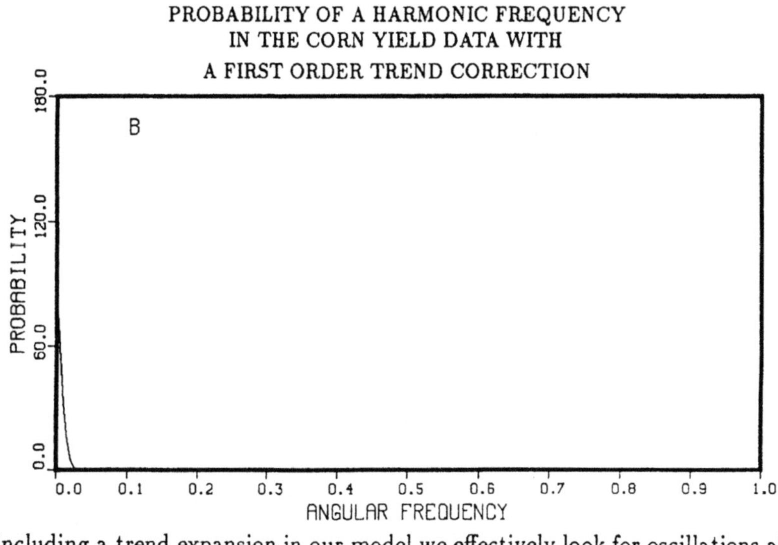

By including a trend expansion in our model we effectively look for oscillations about a trend. This is not the same as detrending, because the trend functions and the sine and cosine functions are never orthogonal. The zero order trend (or constant) plus a simple-harmonic-frequency model (A) is dominated by the trend. When we included a linear trend the height of the trend is decreased some, however the trend is still the dominant effect in the analysis.

PROBABILITY OF A HARMONIC FREQUENCY
IN THE CORN YIELD DATA WITH
A SECOND ORDER TREND CORRECTION

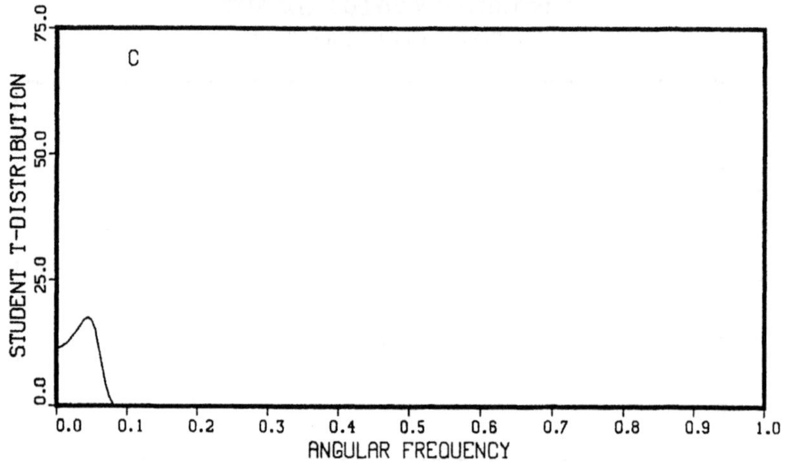

PROBABILITY OF A HARMONIC FREQUENCY
IN THE CORN YIELD DATA WITH
A THIRD ORDER TREND CORRECTION

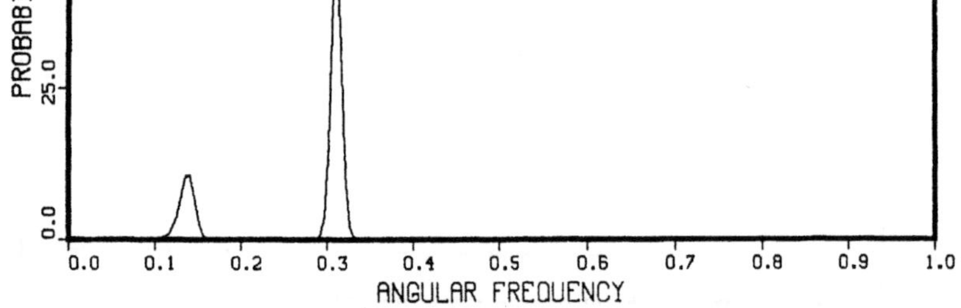

The probability of a single harmonic frequency plus a second-order trend (C) continues to pick out the low frequency trend. However, the level and spread of the marginal posterior probability density is such that the trend has almost been removed. When the probability of a single harmonic frequency plus a third-order trend is computed, the probability density suddenly changes behavior. The frequency near 0.3 is now the dominant feature (D). The trend has not been completely removed; a small artifact persists at low frequencies.

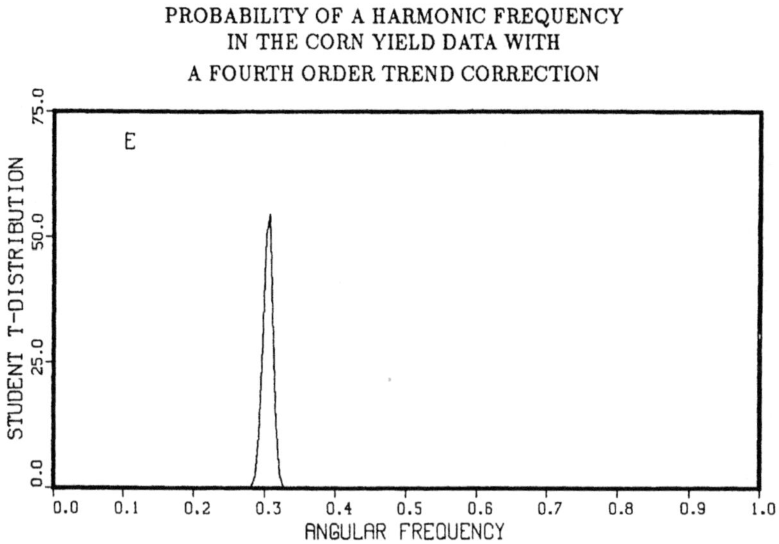

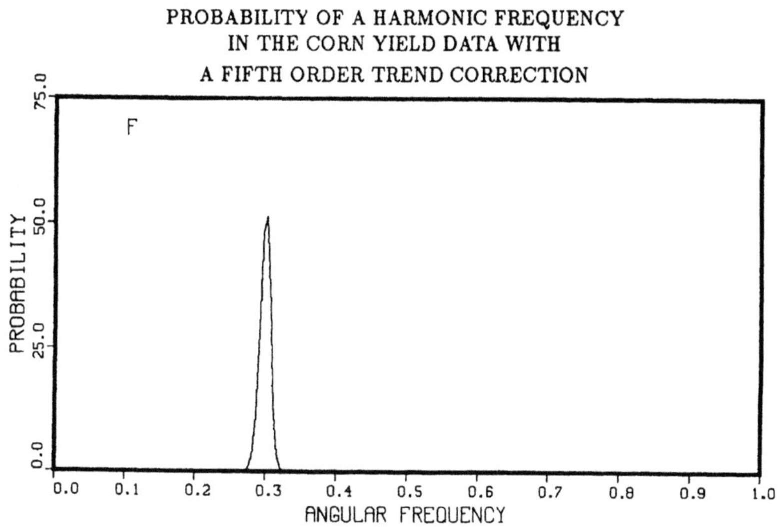

When the probability of a fourth-order trend plus a harmonic frequency is computed the trend is now completely gone and only the frequency at 20 years remains (E). When the expansion order is increased in (F) the frequency estimate is not essentially changed.

PROBABILITY OF A HARMONIC FREQUENCY
IN THE CORN YIELD DATA WITH
A SIXTH ORDER TREND CORRECTION

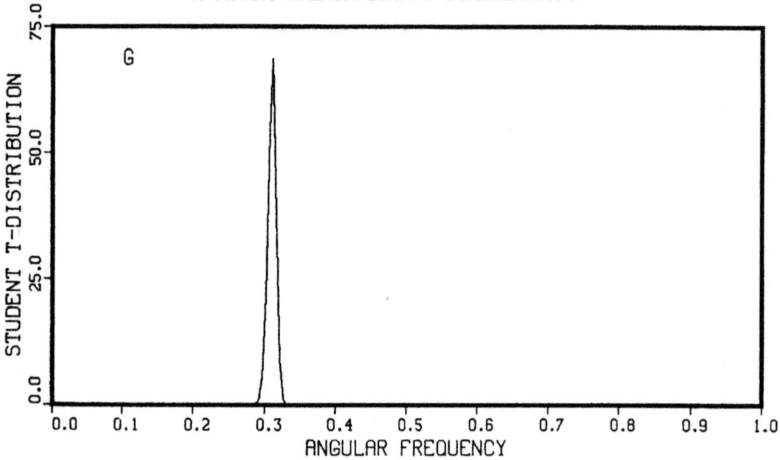

PROBABILITY OF A HARMONIC FREQUENCY
IN THE CORN YIELD DATA WITH
A SEVENTH ORDER TREND CORRECTION

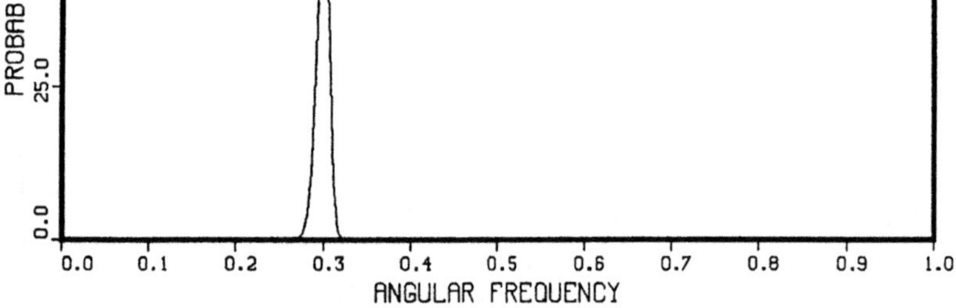

Increasing the expansion order further does not significantly affect the estimated frequency (G) and (H). If the expansion order is increased sufficiently, the expansion will begin to remove the harmonic oscillation; and the posterior probability density will gradually decrease in height.

Figure 7.13: Probability of Two Frequencies After Trend Correction

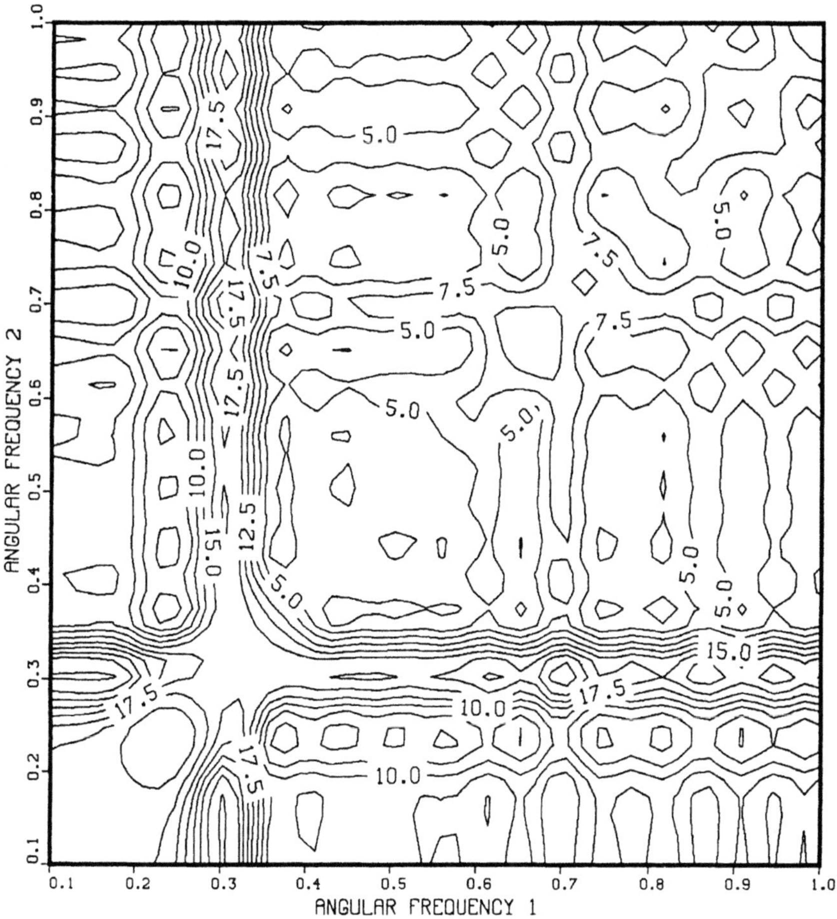

This is the natural logarithm of the probability of two common harmonic frequencies in the crop yield data with a fifth order trend. This type of structure is what one expects from the sufficient statistic when there is only one frequency present. Notice the maximum is located roughly along a vertical and horizontal line at 0.3.

We did not seek to remove the trend from the data, but rather to eliminate its effect from the conclusions.

7.3 Another NMR Example

Now that the tools have been developed we can demonstrate how one can incorporate partial information about a model. In the corn crop example the trend was unknown, so it was expanded in orthonormal polynomials and integrated out of the problem, while we included what partial information we had in the form of the sine and cosine terms. In this NMR example let us assume that the decay function is of interest to us. We would like to determine this function as accurately as possible.

The data we used, Fig. 7.14(A), in this example are one channel of a pure D_2 spectrum [31]. Figure 7.14(B) contains the periodogram for these data. For this demonstration we will use the first $N = 512$ data points because they contain most of the signal.

For D_2, theoretical studies indicate there is a single frequency with decay [32]. Now we expect the signal should have the form

$$f(t) = [B_1 \sin(\omega t) + B_2 \cos(\omega t)]\, D(t),$$

where $D(t)$ is the decay function, and the sine and cosine effectively express what partial information we have about the signal. We will expand the decay function $D(t)$ to obtain

$$f(t) = [B_1 \sin(\omega t) + B_2 \cos(\omega t)] \sum_{j=0}^{r} D_j L_j(t)$$

where D_j are the expansion coefficients for the decay function, B_1 and B_2 are effectively the amplitude and phase of the sinusoidal oscillations, and L_j are the Legendre polynomials with the appropriate change of variables. This model can be rewritten as

$$f(t) = \sum_{j=0}^{r} D_j B_1 \left[L_j(t)[\sin(\omega t) + \frac{B_2}{B_1} \cos(\omega t)] \right].$$

There is an indeterminacy in the overall scale. That is, the amplitude of the sinusoid and the amplitude of the decay $D(t)$ cannot both be determined. One of them is necessarily arbitrary. We chose the amplitude of the sine term to be unity because it effectively eliminates one $\{\omega\}$ parameter from the problem. We have a choice, in this problem, on which parameters are to be removed by integration. We

Figure 7.14: A Second NMR Example - Decay Envelope Extraction

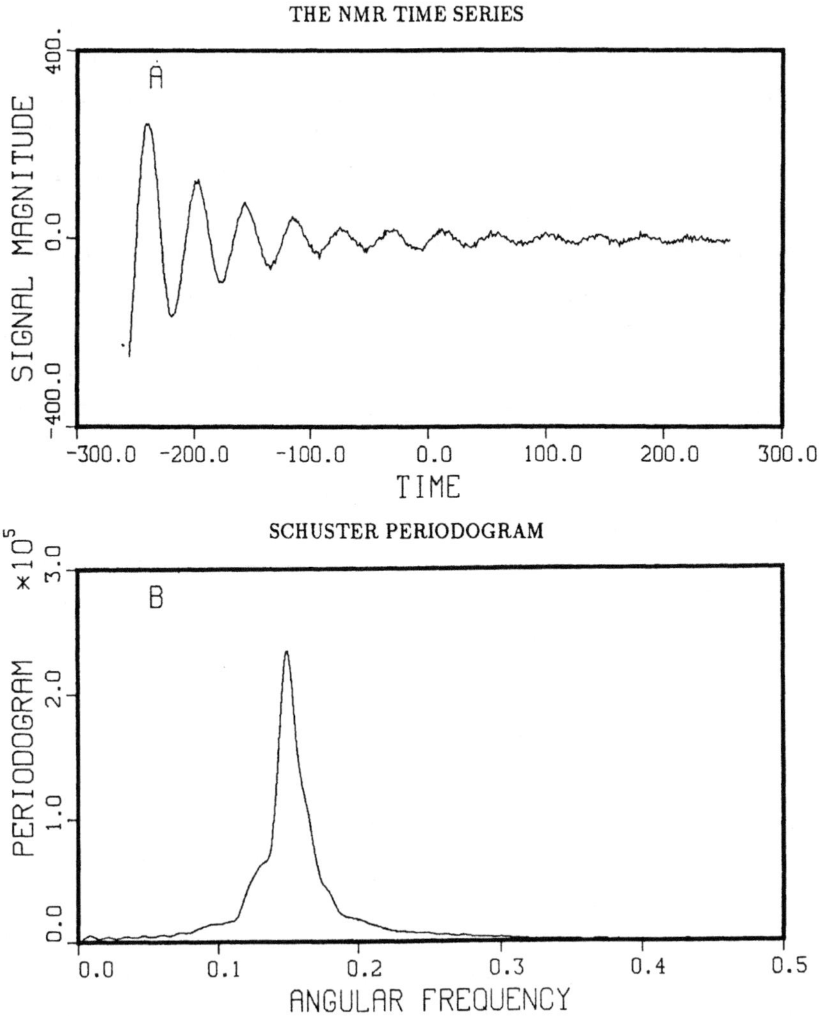

These NMR data (A) are a free-induction decay for a D_2 sample. The sample was excited using a 55MHz pulse and the signal detected using a mixer-demodulator. We used 512 data samples to compute the periodogram (B). We would like to use probability theory to obtain an estimate of the decay function while incorporating what little we know about the oscillations.

chose to eliminate $\{D_j B_1\}$ because there are many more of them, even though they are really the parameters of interest.

When we eliminate a parameter from the problem, it does not mean that it cannot be estimated. In fact, we can always calculate the parameters $\{D_j B_1\}$ from the linear relations between models, Eq. (4.2). For this problem it is simpler to search for the maximum of the probability distribution as a function of frequency ω and the ratio B_1/B_2, and then use Eq. (4.2) to compute the expansion coefficients D_j. If we choose to eliminate the amplitudes of the sine and cosine terms, then we must search for the maximum of the probability distribution as a function of the expansion parameters; there could be a large number of these.

We must again set the expansion order r; here we have plenty of data so in principle we could take r to be large. However, unless the decay is rapidly varying we would expect a moderate expansion of perhaps 5th to 10th order to be more than adequate. In the examples given here we set the expansion order to 10. We solved the problem also with the expansion order set to 5, and the results were effectively identical to the tenth order expansion.

To solve this problem we again used the computer code in Appendix E, and the "pattern" search routine discussed earlier. We located the maximum of the two dimensional "Student t-distribution," Eq. (3.17), and used the procedure given in Chapter 4, Eqs. (4.9) through (4.14), to estimate the standard deviation of the parameters. We find these to be

$$(\omega)_{\text{est}} = 0.14976 \pm 2 \times 10^{-5}$$

$$\left(\frac{B_2}{B_1}\right)_{\text{est}} = -.475 \pm 5 \times 10^{-3}$$

at two standard deviations. The variance of these data was $\overline{d^2} = 2902$, the estimated noise variance $(\sigma^2)_{\text{est}} \approx 27.1$, and the signal-to-noise ratio was 23.3.

After locating the maximum of the probability density we used the linear relations (4.2) between the orthonormal model and the nonorthonormal model to compute the expansion coefficients. We set the scale by requiring the decay function and the reconstructed model function to touch at one point near the global maximum. We have plotted the data and the estimated decay function, Fig. 7.15(A). In Fig. 7.15(B) we have a close up of the data, the decay function, and the reconstructed signal.

It is apparent from this plot that the decay is not Lorentzian or there is a second very small frequency present in the data. The decay function drops rapidly and then begins to oscillate. This is a real effect and is not an artifact of the procedure we are

Figure 7.15: How Does an NMR Signal Decay?

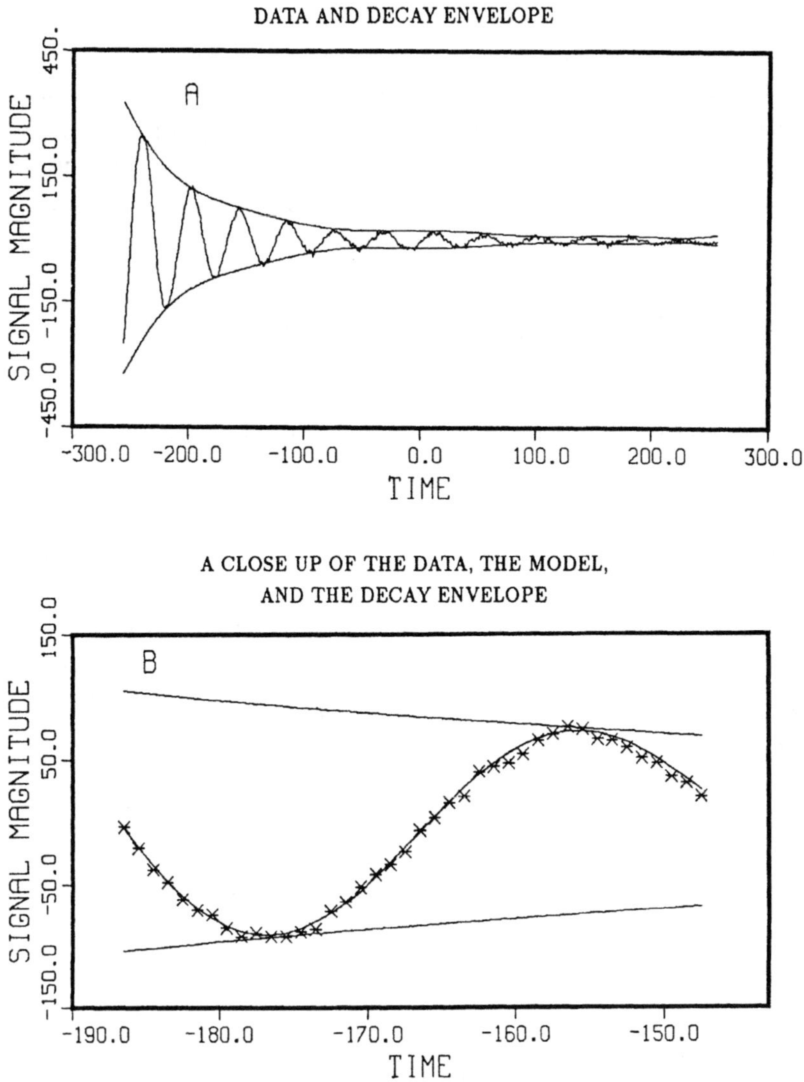

The decay function in (A) comes down smoothly and then begins to oscillate. This is a real effect, and is not an artifact of the analysis. In (B) we have plotted a blow up of the data, the predicted signal, and the decay function.

using. There are two possible interpretations: there could be a second small signal which is beating against the primary signal, or the inhomogeneous magnetic field could be causing it. When a sample is placed in a magnetic field each individual dipole in the field precesses at a well defined rate proportional to the local magnetic field. When the field is inhomogeneous (badly shimmed) a sample will resonate with an entire spectrum of frequencies around the principal frequency. Typically this distribution will manifest itself microscopically as a broadening or perhaps a splitting in lines: they become doubles and that is what we see here as this small oscillation. If we were to go back and look at this resonance very carefully we would find a second very small peak.

7.4 Wolf's Relative Sunspot Numbers

In 1848 Rudolph Wolf introduced the relative sunspot numbers as a measure of solar activity. These numbers, defined earlier, are available as yearly averages since 1700 – Fig. 2.1(A). The importance of these numbers is primarily because they are the longest available quantitative index of the sun's internal activity. The most prominent feature in these numbers is the 11.04 year cycle mentioned earlier. In addition to this cycle a number of others have been reported including cycles of 180, 90, 45, and a 22 years as well as a number of others [37], [38]. We will apply probability theory to these numbers to see what can be learned. We must stress that in what follows we do not know what the "true" model is, but can only examine a number of different possibilities. We begin by trying to determine the approximate number of degrees of freedom any reasonable model of these numbers should have.

7.4.1 Orthogonal Expansion of the Relative Sunspot Numbers

We can get a better understanding of the sunspot numbers if we simply expand these numbers in orthogonal vectors, and allow Eqs. (5.1) and (5.9) to indicate the number of expansion vectors needed to represent the data. This slight variant of the discrete Fourier transform will serve several useful purposes: it will give us an indication of the complexity of the data set, and it will indicate the noise level.

For this simple expansion we used sines and cosines. We generated the cosine vectors using

$$H_j(t_i) = \frac{1}{\sqrt{c_j}} \cos\left(\frac{\pi j t_i}{N}\right)$$

$$c_j \equiv \sum_{i=1}^{N} \cos^2\left(\frac{\pi j t_i}{N}\right)$$

and the sine vectors using

$$H_k(t_i) = \frac{1}{\sqrt{s_j}} \sin\left(\frac{\pi k t_i}{N}\right)$$

$$s_j \equiv \sum_{i=1}^{N} \sin^2\left(\frac{\pi k t_i}{N}\right)$$

where $0 \leq k \leq N/2$ for the cosine components and $1 \leq k \leq N/2$ for the sine components. There are a total of 285 expansion vectors, and for this problem the time increments are one year. Next we computed h_k: the dot product between the data and the expansion vectors. Both the sine and cosine dot products were then squared and sorted into decreasing order.

From these ordered projections we could then easily compute the probability of the expansion order E. For this problem this is essentially the posterior probability Eq. (5.9) with $r = 0$ and the terms associated with the $\{\omega\}$ set equal to 1. Because we are using an orthonormal expansion the Jacobian is unity. This simplifies Eq. (5.9) somewhat; we have

$$P(E|D,I) = \Gamma\left(\frac{E}{2}\right)\Gamma\left(\frac{N-E}{2}\right)\left[\frac{E(\overline{h^2})_E}{2}\right]^{-\frac{E}{2}}\left[\frac{N-E}{2}\langle\sigma^2\rangle\right]^{\frac{E-N}{2}},$$

where $(\overline{h^2})_E$ is the sufficient statistic computed with the E largest orthonormal projections. Figure 7.16 is a plot of the posterior probability of the model as a function of expansion order E. One can see from the plot that there is a peak in the probability around 90, and if one wants to be certain that all of the systematic component has been expanded, then the expansion order must be taken to be approximately 100. The estimated signal-to-noise ratio of these data is approximately 11.5, and the estimated standard deviation is about 5.

An orthogonal expansion of the data is about the worst model one could pick, in the sense of having the largest number of degrees of freedom. If we were to produce a model that reduced the total number of degrees of freedom by a factor 3 we would still have over thirty. For a simple harmonic frequency model, that would be 10 to 14 total frequencies. There are 286 data values, and the main period of roughly 11

Figure 7.16: The Probability of the Expansion Order

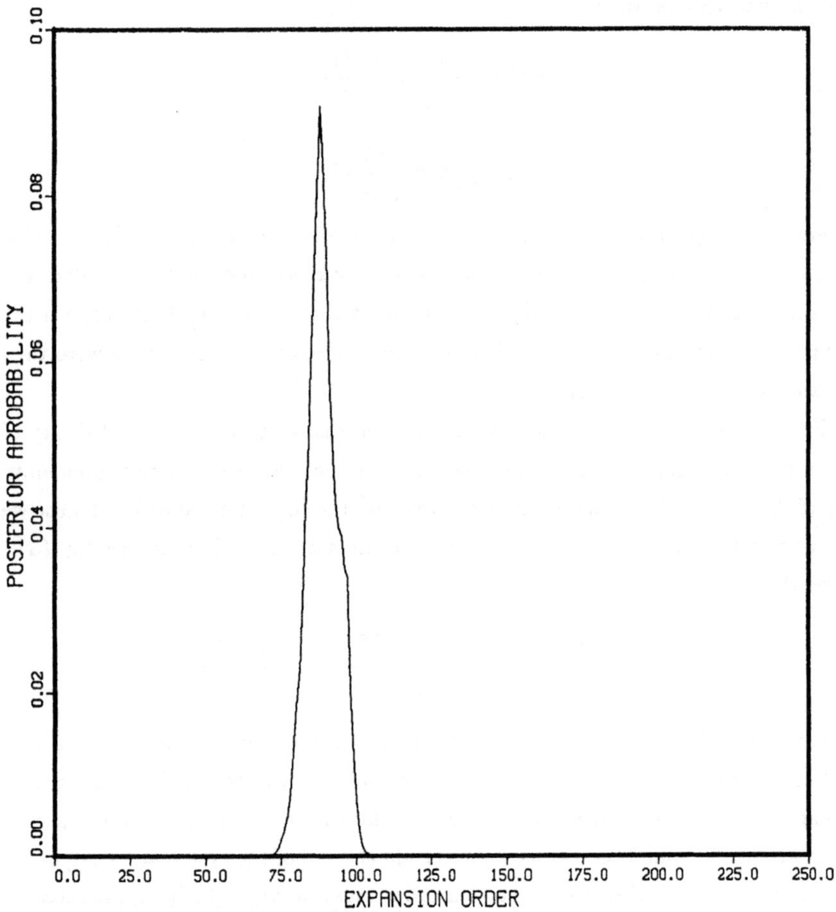

We expanded the Wolf sunspot numbers on orthonormal vectors and then used Eq. (5.9) to decide when to stop the expansion. This probability density indicates that the sunspot numbers are an extremely complex data set needing approximately 100 degrees of freedom to represent them.

years; that gives 26 cycles in the record. If each period has a unique amplitude, that still leaves approximately six to ten degrees of freedom to describe the shape of the oscillation. The implication of this is that Wolf's numbers are intrinsically extremely complicated, and no simple model for these numbers is going to prove possible. We will investigate them using a number of relatively simple models to see what can be learned.

7.4.2 Harmonic Analysis of the Relative Sunspot Numbers

The second model we will investigate is the multiple harmonic frequency model. There are three degrees of freedom for each frequency, and with 100 degrees of freedom in the data, there is no chance of finding all of the structure in them. We will content ourselves with finding the first few frequencies and seeing how the results compare with the orthogonal expansion. Many writers have performed a harmonic analysis on these numbers. We will compare our results to those obtained recently by Sonett [38] and Bracewell [39]. The analysis done by Sonett concentrated on determining the spectrum of the relative sunspot numbers. He used the Burg [36] algorithm. This routine is extremely sensitive to the frequencies. In addition to finding the frequencies, this routine will sometimes shift the location of the predicted frequency, and it estimates a spectral density (a power normalized probability distribution), not the power carried in a line. Consequently, no accurate determination of the power carried by these lines has been done. As explained by Jaynes [40], the Burg algorithm yields the optimal solution to a certain well-defined problem. But in practice it is used in some very different problems for which it is not optimal (although still useful). We will use probability theory to estimate the frequencies, their accuracy, the amplitudes, the phases, as well as the power carried by each line.

Again, we plot the log of the probability of a single harmonic frequency plus a constant, Fig. 7.17(A). In this study, we include a constant and allow probability theory to remove it the correct way, instead of subtracting the average from the data as was done in Chapter 2. We do this to see if this theoretically correct way of eliminating a constant will make any difference in the evidence for frequencies. Thus we plot the log of the marginal posterior probability Eq. (3.17) using

$$f(t) = B_1 + B_2 \cos \omega t + B_3 \sin \omega t$$

Figure 7.17: Adding a Constant to the Model

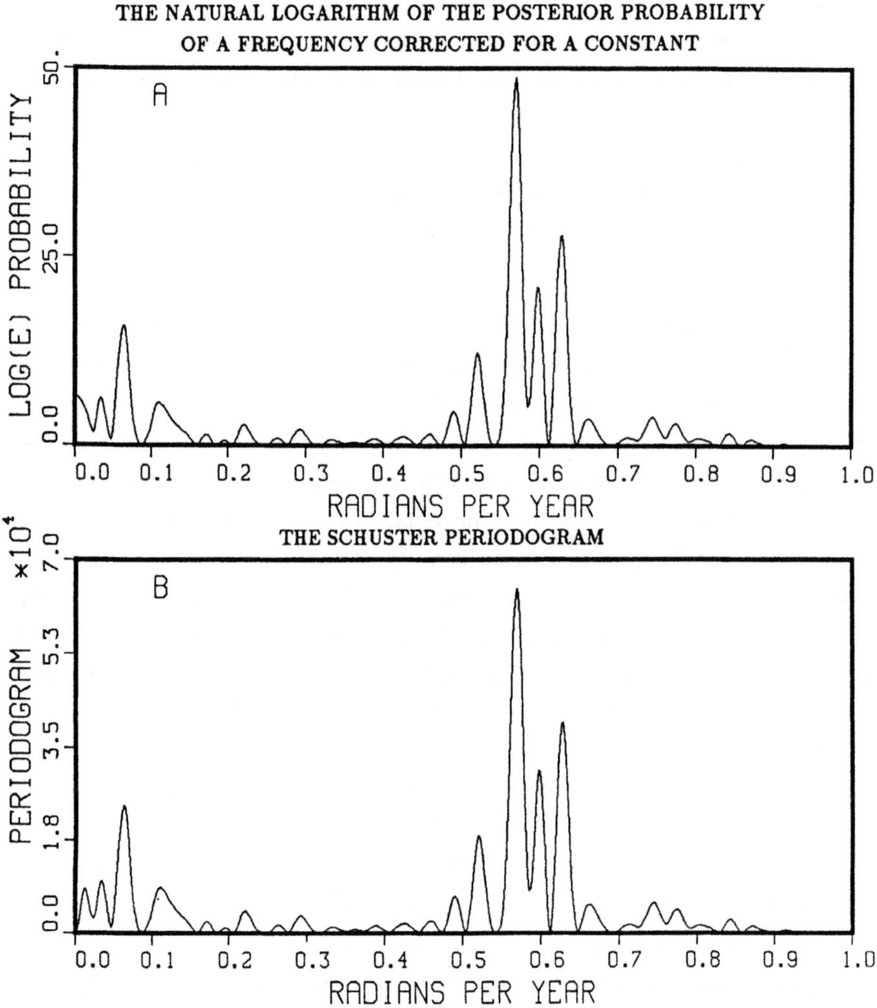

The $\log_e$ of the marginal posterior probability of a single harmonic frequency plus a constant (A), and the periodogram (B) are almost identical. The periodogram is related to the posterior probability when σ^2 is known; for a data set with zero mean the periodogram must go to zero at zero frequency. The low frequency peak near zero in (B) is caused by subtracting the average from the data. The $\log_e$ of the marginal posterior probability of a single harmonic frequency plus a constant will go to zero at zero only if there is no evidence of a constant component in the data. Thus (A) does not indicate the presence of a spurious low frequency peak, only a constant.

as the model. The periodogram, Fig. 7.17(B), is a sufficient statistic for a single harmonic frequency if and only if the time series has zero mean. Under these conditions the periodogram must go to zero at $\omega = 0$. But this is the only difference visible; in the periodogram, Fig. 7.17(B), the low frequency peak near zero is a spurious effect due to subtracting the average value from the data. Probability analysis using a simple harmonic frequency plus a constant does not show any evidence for this period, Fig. 7.17(A).

Next we applied the general procedure for finding multiple frequencies. We started with the single frequency which best described the data, then computed the residuals and looked to see if there was evidence for additional frequencies in the residuals. The initial estimate from the residuals was then used in a two-frequency model. We continued this process until we had a nine-frequency model. Next we computed the standard deviation using the procedure developed in Chapter 4, Eqs. (4.9) through (4.14). Last, we used the linear relations between the models, Eq. (4.2), to compute the nonorthonormal amplitudes as well as their second moments. These are summarized as in Table . With these nine frequencies and one constant, the estimated standard deviation of the noise is $(\sigma)_{\text{est}} = 15$, and the signal-to-noise ratio is 14. The constant term had a value of 46.

We have plotted these nine frequencies as normalized Gaussians, Fig. 7.18(A), to get a better understanding of their determination. We plot in Fig. 7.18(B) an approximation to the line spectral density obtained by normalizing Fig. 7.18(A) to the appropriate power level. The dotted line on this plot is the periodogram normalized to the highest value in the power spectral density. This plot brings home the fact that when the frequencies are close, the periodogram is not even approximately a sufficient statistic for estimating multiple harmonic frequencies. At least one of the frequencies found by the nine-frequency model occurs right at a minimum of the periodogram. Also notice that the normalized power is more or less in fair agreement with the periodogram when the frequencies are well separated. That is because, for a simple harmonic frequency, the peak of the periodogram is indeed a good estimate of the energy carried in that line.

In Fig. 7.19(A), we can plot the simulated sunspot series. We have repeated the plot of the sunspot numbers, Fig. 7.19(B), for comparison. This simple nine-frequency model reproduces most of the features of the sunspot numbers, but there is still something missing from the model. In particular the data values drop uniformly to zero at the minima. This behavior is not repeated in the nine-frequency model.

Table 7.1: The Nine Largest Sinusoidal Components in the Sunspot Numbers

$\langle \hat{f} \rangle_{\text{est}}$	$\langle B_1 \rangle$	$\langle B_2 \rangle$	$\sqrt{B_1^2 + B_2^2}$
11.02 $\pm$ 0.01 years	-35	4.5	35
10.73 $\pm$ 0.03 years	1.0	19	19
9.98 $\pm$ 0.01 years	15	-10	18
88.08 $\pm$ 0.02 years	2.9	-17	17
53.96 $\pm$ 0.02 years	-10	-13	16
11.85 $\pm$ 0.01 years	-14	-2.2	14
48.44 $\pm$ 0.04 years	-9.8	-3.1	10
8.39 $\pm$ 0.03 years	-5.4	6.9	9
13.16 $\pm$ 0.03 years	4.7	-6.6	8

The first column is the frequency with an estimate of the variance of the posterior probability density; the second and third columns are amplitudes of the cosine and sine components and the last column is the magnitude of the signal. There are any number of effects in these data, but the largest is the 11 year cycle. We demonstrated in Section 6.1.4, page 76 that when the oscillations are nonharmonic the single frequency model can have spurious multiple peaks. It is only the largest peak in the marginal posterior probability density of a single harmonic frequency plus a constant that is indicative of the oscillation in the data. If we use a multiple harmonic frequency model, as we did here, probability theory will interpret these spurious peaks as frequencies. Probably all of the effects other than the 11 year cycle are artifacts of not knowing the correct model, which presumably involves nonsinusoidal and nonstationary oscillations.

Figure 7.18: The Posterior Probability of Nine Frequencies

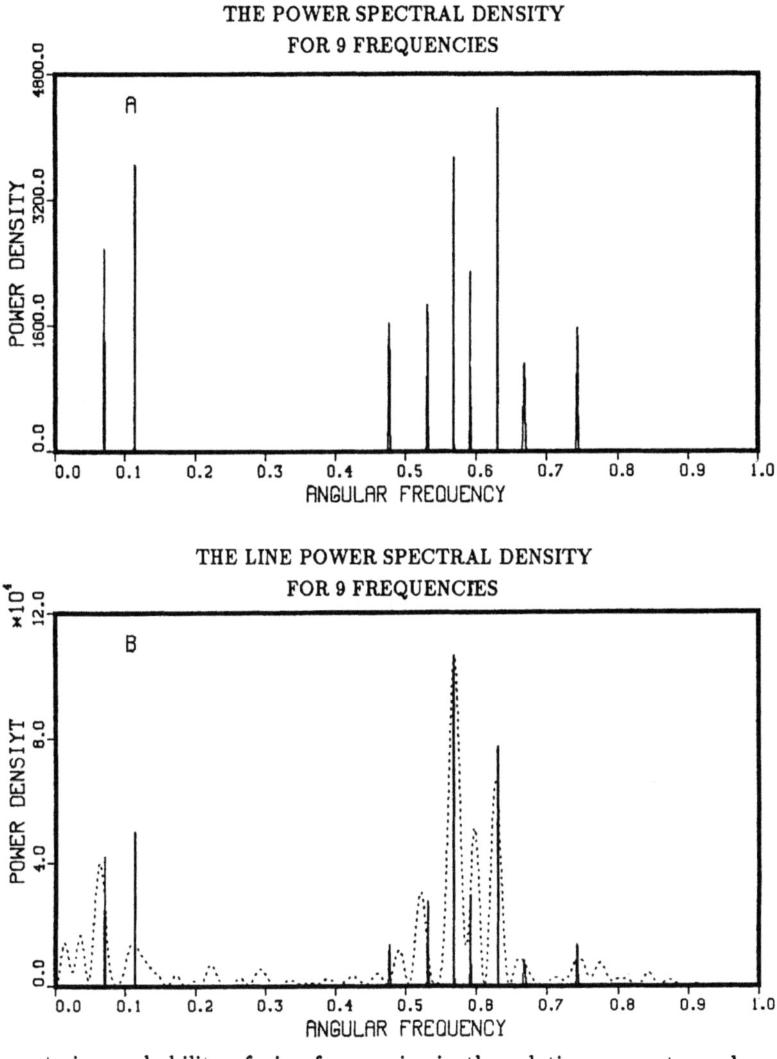

The posterior probability of nine frequencies in the relative sunspot numbers (A), has nine well resolved peaks. In (B) we have a line spectral density. The peak value of the periodogram is an accurate estimate of the energy carried in a line as long as there is only one isolated frequency present.

Figure 7.19: The Predicted Sunspot Series

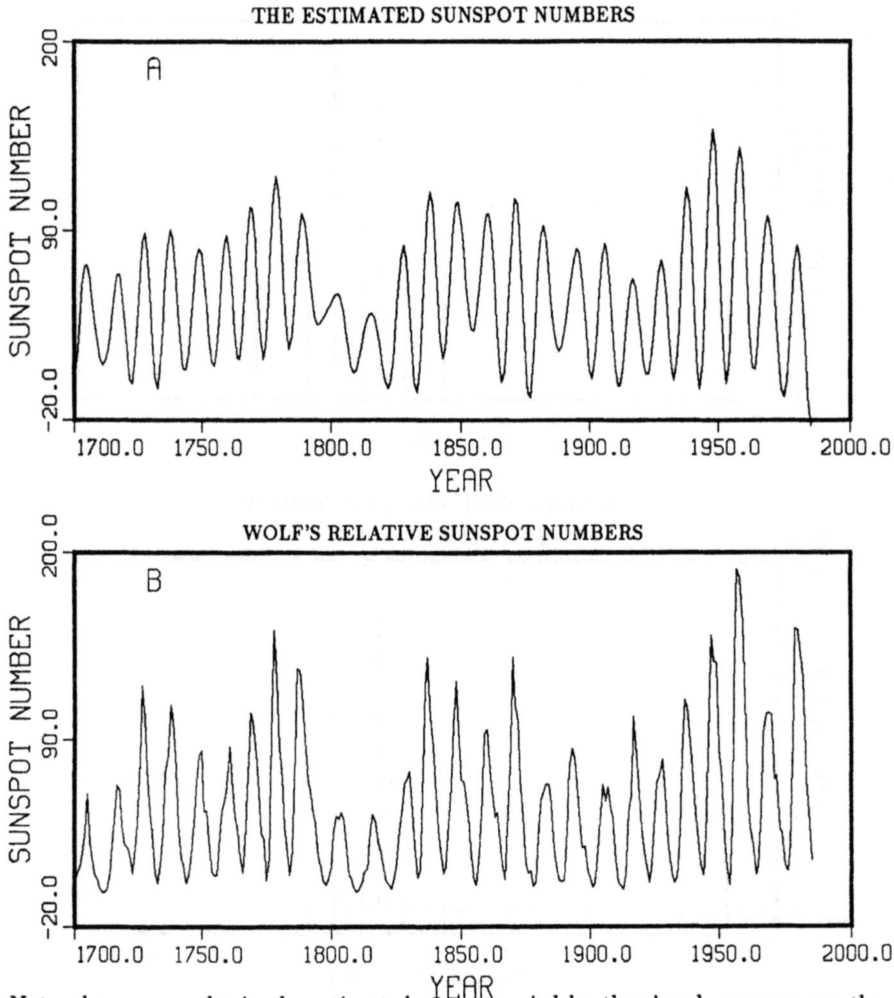

Not only can one obtain the estimated power carried by the signal, one can use the amplitudes to plot what probability theory has estimated to be the signal (A). We have included the relative sunspot numbers (B), for easy comparison.

Also, the data have sharper peaks than troughs, while our sinusoidal model, of course, does not. This is, as has been noted before, evidence of some kind of "rectification" process. A better model could easily reproduce these effects.

7.4.3 The Sunspot Numbers in Terms of Harmonically Related Frequencies

We used a harmonic model on the sunspot numbers so that a simple comparison to a model proposed by C. P. Sonett [38] could be done. He attempted to explain the sunspot numbers in terms of harmonic frequencies; 180, 90, and 45 are examples of harmonically related frequencies. In 1982, Sonett [38] published a paper in which the sunspot number spectrum was to be explained using

$$f(t) = [1 + \alpha \cos(\omega_m t)][\cos(\omega_c t) + \Delta]^2$$

as a model. Sonett's estimate of the magnetic cycle frequency ω_m is approximately 90 years, and his estimate of the solar cycle frequency ω_c is 22 years. The rectification effect is present here.

This model is written in a deceptively simple form and a number of constants (phases and amplitudes) have been suppressed. We propose to apply probability theory using this model to estimate ω_c and ω_m. To do this, we first square the term in brackets and then use trigonometric identities to reduce this model to a form in which probability theory can readily estimate the amplitudes and phases:

$$\begin{aligned}
f(t) = B_1 &+ B_2 \cos([\omega_m]t) &&+ B_3 \sin([\omega_m]t) \\
&+ B_4 \cos([2\omega_m]t) &&+ B_5 \sin([2\omega_m]t) \\
&+ B_6 \cos([\omega_c - 2\omega_m]t) &&+ B_7 \sin([\omega_c - 2\omega_m]t) \\
&+ B_8 \cos([\omega_c - \omega_m]t) &&+ B_9 \sin([\omega_c - \omega_m]t) \\
&+ B_{10} \cos([\omega_c]t) &&+ B_{11} \sin([\omega_c]t) \\
&+ B_{12} \cos([\omega_c + \omega_m]t) &&+ B_{13} \sin([\omega_c + \omega_m]t) \\
&+ B_{14} \cos([\omega_c + 2\omega_m]t) &&+ B_{15} \sin([\omega_c + 2\omega_m]t) \\
&+ B_{16} \cos([2\omega_c - 2\omega_m]t) &&+ B_{17} \sin([2\omega_c - 2\omega_m]t) \\
&+ B_{18} \cos([2\omega_c - \omega_m]t) &&+ B_{19} \sin([2\omega_c - \omega_m]t) \\
&+ B_{20} \cos([2\omega_c]t) &&+ B_{21} \sin([2\omega_c]t) \\
&+ B_{22} \cos([2\omega_c + \omega_m]t) &&+ B_{23} \sin([2\omega_c + \omega_m]t) \\
&+ B_{24} \cos([2\omega_c + 2\omega_m]t) &&+ B_{25} \sin([2\omega_c + 2\omega_m]t).
\end{aligned}$$

Now Sonett specifies the amplitudes of these, but not the phases [38]. We will take a more general approach and not constrain these amplitudes. We will simply allow probability theory to pick the amplitudes and phases which fit the data best. Thus any result we find will have the Sonett frequencies ω_m and ω_c, but the amplitudes and phases will be chosen in a way that will fit the data at least as well as does the Sonett model – possibly somewhat better. After integrating out the amplitudes we have only two parameters to determine, ω_c and ω_m.

We located the maximum of the posterior probability density using the computer code in Appendix E and the pattern search routine. The "best" estimated value for ω_c (in years) is approximately 21.0 years, and for ω_m approximately 643 years. The values for these parameters given by Sonett are $\omega_c = 22$ years and $76 < \omega_m < 108$ years with a mean value of $\omega_m \approx 89$ years. Our probability analysis estimates the values of ω_c to be about the same, and ω_m to be substantially different, from those given by Sonett. The most indicative value is the estimated standard deviation for this model: $\sigma_{\text{Sonett}} = 25.5$ years. By this criterion, this model is no better than a four-frequency model. Considering that a four-frequency model has 15 degrees of freedom compared to 29 for this model, we can all but exclude harmonically related frequencies as a possible explanation of the sunspot numbers. Of course, these conclusions refer only to an analysis of the entire run of data; if we considered the first century of the record to be unreliable and analyzed only the more recent data, a different conclusion might result.

7.4.4 Chirp in the Sunspot Numbers

We have so far investigated two variations of harmonic analysis of the relative sunspot numbers. Let us proceed to investigate a more complex case to see whether there is more structure in the relative sunspot numbers than just simple periodic behavior. These data have been looked at from this standpoint at least once before. Bracewell [39] has analyzed these numbers to determine whether they could have a time-dependent "instantaneous phase". The model used by Bracewell can be written as

$$f(t) = B_1 + \text{Re}\,[E(t)\exp(i\phi(t) + i\omega_{11}t)]$$

where B_1 is a constant term in the data, $E(t)$ is a time varying magnitude of the oscillation, $\phi(t)$ is the "instantaneous phase", and ω_{11} is the 11 year cycle.

This model does not incorporate any prior information into the problem. It is so general that any function can be written in this form. Nevertheless, the idea that the phase $\phi(t)$ could be varying slowly with time is interesting and worth investigating.

An "instantaneous phase" in the notation we have been using is a chirp. Let $\phi(t)$ stand for the phase of the signal, and ω its frequency. Then we may Taylor expand $\phi(t)$ around $t = 0$ to obtain

$$\omega t + \phi(t) \approx \phi_0 + \omega t + \frac{\phi''}{2} t^2 + \cdots,$$

where we have assumed $\phi'(t) = 0$. If this were not so then ω is not the frequency as presumed here. The Bracewell model can then be approximated as

$$f(t) = B_1 + E(t)[\cos(\omega t + \alpha t^2) + B_2 \sin(\omega t + \alpha t^2)].$$

Thus, to second order, the Bracewell model is just a chirped frequency with a time varying envelope.

We can investigate the possibility of a chirped signal using

$$f(t) = C_1 + C_2 \cos(\omega t + \alpha t^2) + C_3 \sin(\omega t + \alpha t^2)$$

as the model, where α is the chirp rate, C_1 is a constant component, ω is the frequency of the oscillation, and C_2 and C_3 are effectively the amplitude and phase of the oscillation. This model is not a substitute for the Bracewell model. Instead this model is designed to allow us to investigate the possibility that the sunspot numbers contain evidence of a chirp, or "instantaneous phase" in the Bracewell terminology.

A plot of the log of the "Student t-distribution" using this model is the proper statistic to look for chirp. However, we now have two parameters to plot, not one. In Fig.7.20 we have constructed a contour plot around the 11 year cycle. We expect this plot to have a peak near the location of a frequency. It will be centered at zero chirp rate if there is no evidence for chirp, and at some nonzero value when there is evidence for chirp. Notice, that along the line $\alpha = 0$ this "Student t-distribution" is just the simple harmonic probability distribution studied earlier, see Fig. 2.1(A). As with the Fourier transform if there are multiple well separated chirped frequencies (with small chirp rates) then we expect there to be multiple peaks in Fig. 7.20.

There are indeed a number of peaks; the single largest point on the plot is located off the $\alpha = 0$ axis. The data contain evidence for chirp. The low frequencies also show evidence for chirp. To the extent that the Bracewell "instantaneous phase" may

Figure 7.20: Chirp in the Sunspot Numbers?

LOG$_{10}$ PROBABILITY OF A CHIRPED FREQUENCY

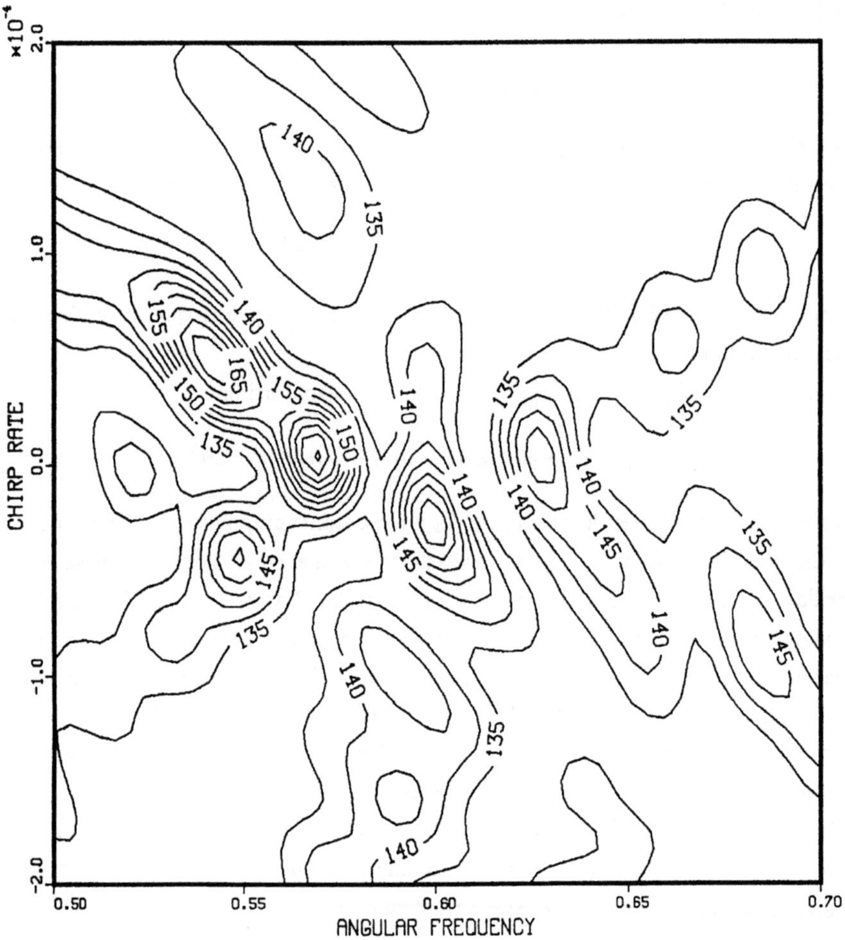

To check for chirp we take $f(t) = A_1 + A_2\cos(\omega t + \alpha t^2) + A_3\sin(\omega t + \alpha t^2)$ as the model. After integrating out the nuisance parameters, the posterior probability is a function of two variables, the frequency ω and the chirp rate α. We then plotted the $\log_e$ of the posterior probability. The single highest peak is located at a positive value of α: there is evidence of chirp.

be considered as a chirp, we must agree with him: there is evidence in these data for chirp.

In light of this discussion, exactly what these numbers represent and exactly what is going on inside the sun to produce them must be reconsidered. The orthogonal expansion on these numbers indicates that the complexity of these numbers is immense and no simple model will suffice to explain them. Given the total number of degrees of freedom it is likely that every cycle has a unique amplitude and a complex nonsinusoidal shape. In other words, different sunspot cycles are about as complicated in structure as are different business cycles in economic data. If that is true, the only frequency in these data of any relevance is probably the 11 year cycle; the other indications of frequency are just effects of the nonharmonic oscillation. Again, had we analyzed only the more recent data, the conclusions might have been different. Certainly we have not answered any real questions about what is going on; indeed that was not our intention. Instead we have shown how use of probability theory for data analysis can facilitate future research by testing various hypotheses more sensitively than could the traditional intuitive *ad hoc* procedures.

7.5 Multiple Measurements

The traditional way to analyze multiple (i.e. multi-channel) measurements is to average the data, and then analyze the averaged data. The hoped-for improvement in the parameter estimates is the standard $\sqrt{n}$ rule. To derive this rule one must assume that the signal and the noise variance, are the same in every data set, and that the noise samples were uncorrelated. Unfortunately, the conditions under which averaging works at its theoretical best are almost never realized in real experiments. Specifically, all experiments contain some effects which will not average out. These effects can become so significant, that the evidence for the signal can be greater in any one of the data sets that went into the average than it is in the averaged data (we will demonstrate this shortly). There are three main reasons why averaging may fail to give the expected $\sqrt{n}$ improvement in the parameter estimates: the experiment may not be reproducible, the model may incorrect, the noise within different data sets may be correlated.

In real physics experiments, reproducibility depends critically on the electronics repeating itself exactly every time. Of course this never happens; there are always small differences. For example, to repeat an NMR experiment one must bring the sample to a stationary state (this may be far from equilibrium) and then further excite the sample using a high power radio transmitter. In a perfect world, every time one excited the sample it would be with a pulse of exactly the same amplitude and exactly the same shape as before. Of course this never happens; every repetition is a unique experiment having slightly different amplitudes, phases, and noise variance. These slight differences are enough to cause averaging to fail to give the $\sqrt{n}$ improvement when large numbers of data sets are averaged, even when the noise samples are independent.

The second source of systematic error is in our imprecise knowledge of the model. If the signal is exactly the same in each data set, of course the noise is reduced by averaging. Unfortunately, if we do not know the model exactly, then our model is only an approximation. When we fit the model to the data, some of the "true" signal will not be fit. This misfit of the model will be called noise by probability theory. But it is noise that is perfectly correlated in successive data sets, and does not average out. Thus the accuracy estimates will not improve, because the dominant contribution to the estimated noise variance will be the misfit between the model and the data.

If any systematic effect is present, averaging will fail to give the expected improvement; nevertheless, probability theory does not mislead us. We have stressed several times that the estimates one obtains from these procedures are conservative. That is, when the models misfit the data they still give the best estimates of the parameters possible under the circumstances, and yield conservative (wide) error estimates. This suggests that by analyzing each data set separately, and looking for common effects, we might be able to realize better estimates than by averaging. In this section we investigate the effects of multiple measurements and compare the results of a joint analysis (analyzing all of the data) to the analysis of the averaged data. We will do this analysis on a data set that most people would not hesitate to average. This is our first example where we apply Bayesian analysis to data which are not a time series.

The experiment we will consider is a simple diffraction experiment. A mercury vapor lamp was placed in front of a slit, and the light from the lamp passed through the slit and onto a screen. An electronic camera (a Charge Coupled Device – CCD) was placed behind the screen and used to image the intensity variations. The data for this analysis were kindly provided by W. H. Smith [41]. The image consists of a series of light and dark bars typical of such experiments. The pattern for the first row in the CCD is shown in Fig. 7.21(A). Figure 7.21(B) is a plot of the averaged data.

This particular device has 573 pixels in each row and there are 380 rows. Thus there are a total of 380 repeated measurements. The CCD was aligned parallel with the slit, so the image appearing in each row should be identical. In principle the data could be averaged to obtain a $\sqrt{380}$ improvement in the parameter estimates. However, the concerns mentioned earlier are all applicable here. The types of systematic effects that can enter this experiment are numerous, but a few of them are: the camera readout has small systematic variations from one row to the next; there can be intensity variations from the first to last row; and if the alignment of the camera is not perfect, there will be small phase drifts from the first to last row. Nonetheless, when one looks at these data, there is absolutely no reason to believe that averaging should not work.

We begin the analysis by plotting the $\log_{10}$ of the probability of a single harmonic frequency plus a constant. We plot this probability density for the first row of the CCD in Fig. 7.22(A), for the average of the 380 data sets in Fig. 7.22(B), and jointly for all data Fig. 7.22(C). One sees from the average data, Fig. 7.22(B), that there is indeed large evidence for a frequency near 1.6 in dimensionless units. That peak is some 133 orders of magnitude above the noise level. The second thing that one sees is that the peak from the first row of the CCD, Fig. 7.22(A), is some 136 orders of magnitude above the noise: the peak from one row has more evidence for frequencies than the average data! The third plot, Fig. 7.22(C), is from the joint analysis. That peak is some 55,000 orders of magnitude higher than the average data! The implications of this are indeed staggering. If one cannot average data in an experiment as simple as this one, then there are probably no conditions under which averaging is the way to proceed. Because the issues raised by this simple example are so important, we will pause to investigate some of the theoretical implications before proceeding with this example.

7.5.1 The Averaging Rule

To derive the average rule one assumes a signal $f(t)$, and n sets of data $d(t_i)_j$ with noise variance σ^2. The signal $f(t)$ and the noise variance σ^2 are assumed to be the same in every data set. Then the probability that we should obtain a data set $d(t_i)_j$

Figure 7.21: A Simple Diffraction Pattern

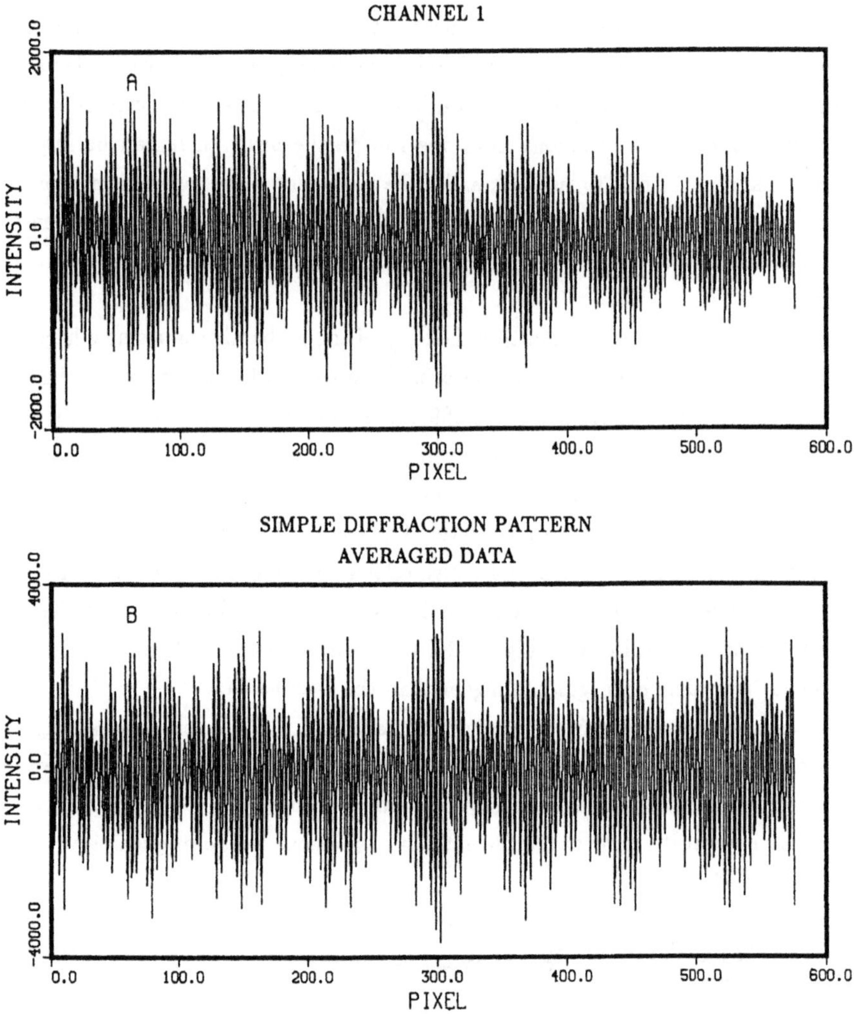

An image was formed by placing a mercury vapor lamp behind a slit and allowing its light to shine through a slit and onto a screen, the CCD imaged the screen. The first row from the CCD is shown in (A). This particular CCD was 573 by 380 pixels, so there are 380 channels. The averaged data is shown in (B). The expected improvement in resolution is $\sqrt{380}$. However, if there are systematic errors in the data, the actual improvement realized will be less.

Figure 7.22: Log_{10} Probability of a Single Harmonic Frequency

LOG_{10} PROBABILITY
IN CHANNEL 1

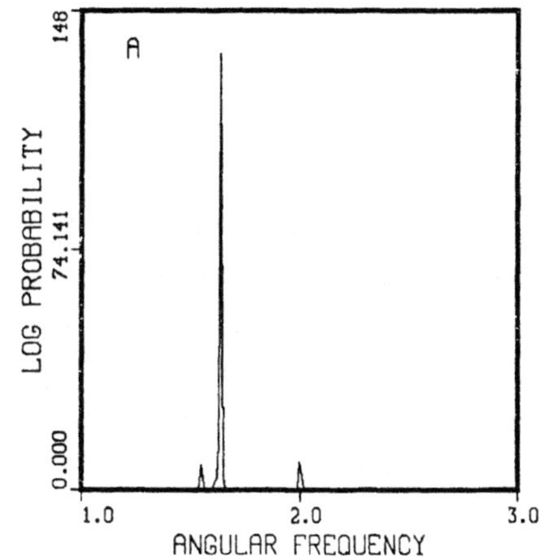

LOG_{10} PROBABILITY
IN THE AVERAGED DATA

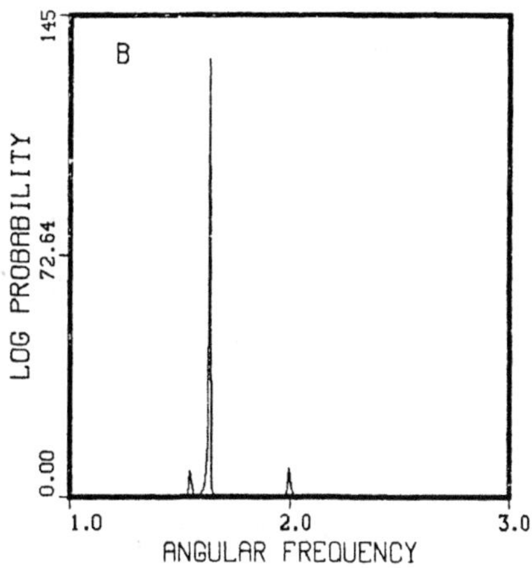

LOG_{10} PROBABILITY
IN THE JOINT ANALYSIS

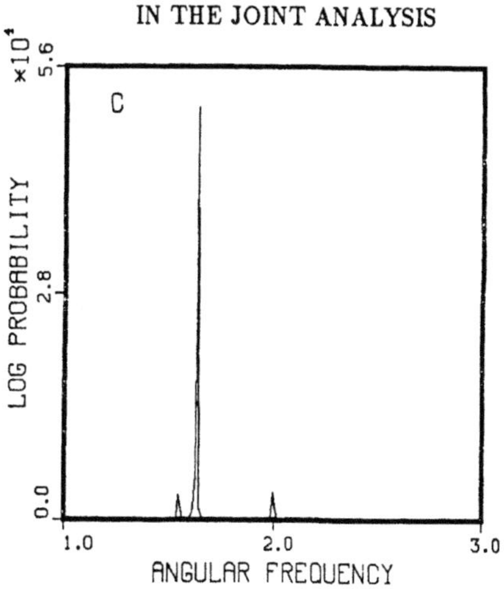

The base 10 logarithm of the probability of a single harmonic frequency in the first row from the CCD shows strong evidence for a frequency (A). The same plot for the average data (B) also has good evidence for a frequency. The base 10 logarithm of the probability of a single harmonic frequency in the joint analysis of the data (C) is some 55,000 orders of magnitude higher.

is given by

$$P(D_j|f,\sigma,I) \propto \exp\left\{-\frac{1}{2\sigma^2}\sum_{i=1}^{N}(d(t_i)_j - f(t_i))^2\right\}.$$

If the noise samples in different channels are independent, the probability that we should obtain all the data sets is just the product of the probabilities that we should obtain any one of the data sets:

$$P(D|f,\sigma,I) \propto \Pi_{j=1}^{n} \exp\left\{-\frac{1}{2\sigma^2}\sum_{i=1}^{N}(d(t_i)_j - f(t_i))^2\right\}.$$

This can be written as

$$P(D|f,\sigma,I) \propto \exp\left\{-\frac{n}{2\sigma^2}\sum_{i=1}^{N}[\overline{d(t_i)^2} - 2\overline{d(t_i)}f(t_i) + f(t_i)^2]\right\},$$

where $\overline{d(t_i)^2}$ is the mean square data value at time t_i, and $\overline{d(t_i)}$ is the mean data value, where "mean" signifies "average over the channels". This is almost a standard model-fitting problem with the data replaced by the average. The procedure is called "Brute Stacking" by geophysicists. The improvement comes from the factor of n multiplying the term in square brackets. We demonstrated that the accuracy estimates are all proportional to the square root of the variance, and here the variance is effectively σ^2/n – this gives the standard $\sqrt{n}$ improvement.

7.5.2 The Resolution Improvement

When multiple data sets are analyzed jointly, how much improvement in the parameter estimates can be expected? The resolution improvement depends on the curvature of the posterior probability density at the maximum. The general rule depends on the model; all we can say is that if all of the data sets have approximately the same evidence in them, then the logarithm of the posterior probability density is n times larger and something like the $\sqrt{n}$ improvement will be realized. We can demonstrate this for the single frequency estimation problem. Then the posterior probability of the frequency, given n repeated measurements and assuming the noise variance σ^2 is the same, is

$$P(\omega|\sigma,D,I) \approx \exp\left\{\sum_{j=1}^{n}\frac{C(\omega)_j}{\sigma^2}\right\} \qquad (7.4)$$

where $C(\omega)_j$ is the Schuster periodogram evaluated for data set j. To obtain the accuracy estimates we expand the exponent about the "true" frequency $\hat{\omega}$ to obtain

$$P(\omega|D,I) \approx \exp\left\{-\sum_{j=1}^{n}\frac{b_j(\hat{\omega}-\omega)^2}{2\sigma^2}\right\},$$

where $b_j = -C_j''$ for the jth data set. If the data contain a single sinusoid such as $\hat{B}\cos(\hat{\omega}t)$, then b is given by Eq. (2.10). This gives the posterior probability density for a simple harmonic frequency when multiple measurement are present as

$$P(\omega|D,I) \approx \exp\left\{-\sum_{j=1}^{n}\frac{N^3\hat{B}_j^2}{96\sigma^2}(\hat{\omega}-\omega)^2\right\}.$$

The accuracy estimate is given by

$$(\omega)_{\text{est}} = \hat{\omega} \pm \sqrt{\frac{48\sigma^2}{N^3 n\overline{\hat{B}^2}}}$$

where $\overline{B^2}$ is the mean square of the true amplitude. If all of the amplitudes are nearly the same height, this is just the standard $\sqrt{n}$ improvement.

The improvement realized is directly related to how well the assumptions in the calculation are met. In the case of averaging, the assumptions are that the amplitude, frequency, phase, and noise variance are the same in every data set. For the example just given we removed the assumption that the amplitudes had to be the same in every data set, consequently, $n\hat{B}^2$ was replaced by $n\overline{\hat{B}^2}$. If we further remove the assumption that the noise variance is the same in every data set, then $n\overline{\hat{B}^2}$ is replaced by $\sum_{j=1}^{n}\hat{B}_j^2/\sigma_j^2$.

When the assumed conditions are not met, the price one pays is in resolution. The procedure described by Eq. (7.2) is more general than averaging in that it allows the amplitudes, phases, and noise variance to be different for each data set and still allows one to look for common effects. When the true amplitudes, phases, and noise variance are the same in every data set this procedure reduces to averaging. Thus Eq. (7.2) represents a more conservative approach than averaging and will realize something approaching the $\sqrt{n}$ improvement under a wider variety of conditions, because it makes fewer assumptions.

7.5.3 Signal Detection

When multiple measurements are present we would like to understand what happens to the joint analysis as we increase the number of measurements. In other words we typically average data when the signal-to-noise ratio is very bad. We do this because we think it allows one to reduce the noise; thus small signals can be detected. But what will happen with the joint analysis? The general answer for the joint analysis again depends on the model function. However, as noted earlier, if the evidence in each data set is roughly the same the sufficient statistic will be n times larger than when only a single measurement is present. Thus the evidence for a signal will build up in a manner similar to averaging. We will demonstrate how the evidence accumulates in the joint analysis using the single frequency model. The posterior probability density of a common harmonic frequency, when multiple measurements are available, is again given by Eq. (7.4). What we would like to know is how high is the peak in Eq. (7.4)? Again taking the data to be

$$d(t)_j = \hat{B}_j \cos(\hat{\omega} t)$$

the peak value of the periodogram is given approximately by

$$C(\hat{\omega})_j \approx \frac{N\hat{B}_j^2}{4}.$$

Assuming each data set has the same number of data values N, the maximum of the posterior probability density will be

$$P(\hat{\omega}|D,I) \propto \exp\left\{\sum_{j=1}^{n} \frac{N\hat{B}_j^2}{4\sigma^2}\right\}.$$

We can simplify this by using

$$\sum_{j=1}^{n} \hat{B}_j^2 = n\overline{B^2}$$

where $\overline{B^2}$ is the mean-square true amplitude. The evidence for a signal increases by the power of the number of data sets:

$$P(\hat{\omega}|D,I) \propto \exp\left\{\frac{nN\overline{B^2}}{4\sigma^2}\right\}.$$

If we allow the variance of the noise to be different in each data set $n\overline{B^2}$ will be replaced by a weighted average $\sum_{j=1}^{n} \hat{B}_j^2/\sigma_j^2$. Again we find the height of the posterior probability density depends directly on the assumptions made in the calculation. In the case of averaging, the log-height of the posterior probability density is n times larger than the height from one data set (provided the assumptions are met). If we relax the assumptions about the amplitude B, we replace B^2 in the average rule by the mean-square true value. If we further relax the assumptions and allow the noise variances to be different, we obtain the weighted average of the true B^2 values. Thus we again have a more conservative procedure that will reduce to give the $\sqrt{n}$ rule when the appropriate conditions are met, under much wider conditions than averaging. In the case of the simple harmonic frequency, doubling the number of data sets is similar to doubling the number of time samples, while keeping the total sampling time fixed. This is not the best way to find a signal (doubling the signal-to-noise works better), but if no other course is available it will build up the probability density by essentially squaring the distribution for each doubling of the number of data sets.

7.5.4 The Distribution of the Sample Estimates

In the CCD example, we had 380 repeated measurements, and the maximum of the posterior probability was some 55,000 orders of magnitude above the noise. Each data set raised the posterior probability approximately 55,000/380 = 144 orders of magnitude. But when the data were averaged, small variations in the amplitude, phases, and variance of the noise caused systematic variations in the data which were nonsinusoidal. Probability theory automatically reduced both the height of the posterior density (i.e. it could not see the signal as well) and reduced the precision of the estimates. In this example the height was reduced from 55,000 to 133, and the accuracy was reduced from 6.8×10^{-8} to 0.00036; the error estimate of the averaged data is 5266 times larger than the estimate from the joint analysis. It thus appears that data averaging (Brute Stacking) is never better than a joint analysis of the data, and it is in general worse. Averaging does, of course, reduce the amount of computation; but with modern computers this is not an important consideration.

In this last example the improvement was very dramatic, but this was real experimental data; perhaps the reason averaging failed was some other effect in the data. We would like to show that the cause was the variation of the signal and the noise variance in the various data sets. To do this we will generate data from the following equation

$$d(t) = B\cos(0.3t + \theta) + e(t). \tag{7.5}$$

We will vary B, θ, and σ from one data set to another. We will then estimate the frequency in the averaged data and in a joint analysis of all the data, and show that these variations will cause averaging to fail to give the $\sqrt{n}$ improvement; while a joint analysis will continue to exhibit the expected improvement.

We generated multiple data sets from Eq. (7.5). Each data set we generated had a different amplitude B, phase θ, and noise variance σ^2. To generate the data we used three Gaussian random numbers with unit variance. We used one as the amplitude B, another as the phase θ, and the third to scale the noise. The noise was generated using a Gaussian random number generator with unit variance. We generated the noise and then multiplied it by the third random number. Using this procedure, the signal will average to zero.

We generated 100 data sets, each containing 512 data values. An example of one such data set is shown in Fig. 7.23(A). The $\log_{10}$ of the probability of a single harmonic frequency is shown in Fig. 7.23(B). We have also displayed the average data Fig. 7.23(C) and the $\log_{10}$ probability of a single frequency. In this particular case averaging has not completely cancelled the signal, however, one measurement, Fig. 7.23(A), has about a 10^9 times more evidence for a signal than the average data, Fig. 7.23(B). We estimated the frequency in the 100 data sets in a joint analysis and in the averaged data. We then selected three new random numbers and repeated this calculation 3000 times.

The results are summarized in Table 7.2. This table contains the actual estimates from a few of the 3000 sets of data analyzed. The second column is the estimated frequency from the average data. The third column is the squared difference between the true frequency and the estimate from the averaged data, (the variance of this estimate). The fourth column is the estimated frequency from the joint analysis, and the fifth column is the squared difference between the true frequency and the estimated frequency from the joint analysis. We averaged all 3000 entries (labeled average at the bottom of the table), and we computed the square root (the standard deviation) estimate for the variance (columns 3 and 5). The estimate from the averaged is a little better than it actually was in these data. When we estimated the frequency we had to give the search routine an initial estimate of the frequency. This locked the search routine onto the correct peak in the periodogram even though there was no clear peak in many of the data sets. This is analogous to estimating the averaged frequency with a strong prior.

For a single data set with unit signal-to-noise the "best" estimated frequency should be ± 0.0006 radians per step; the estimated standard deviation of the averaged data is about a factor of 2 larger than what one would obtain from one data set. Thus averaging has destroyed evidence in the data: any one data set contains more evidence for frequencies than the averaged data. If averaging were working

Figure 7.23: Example - Multiple Measurements

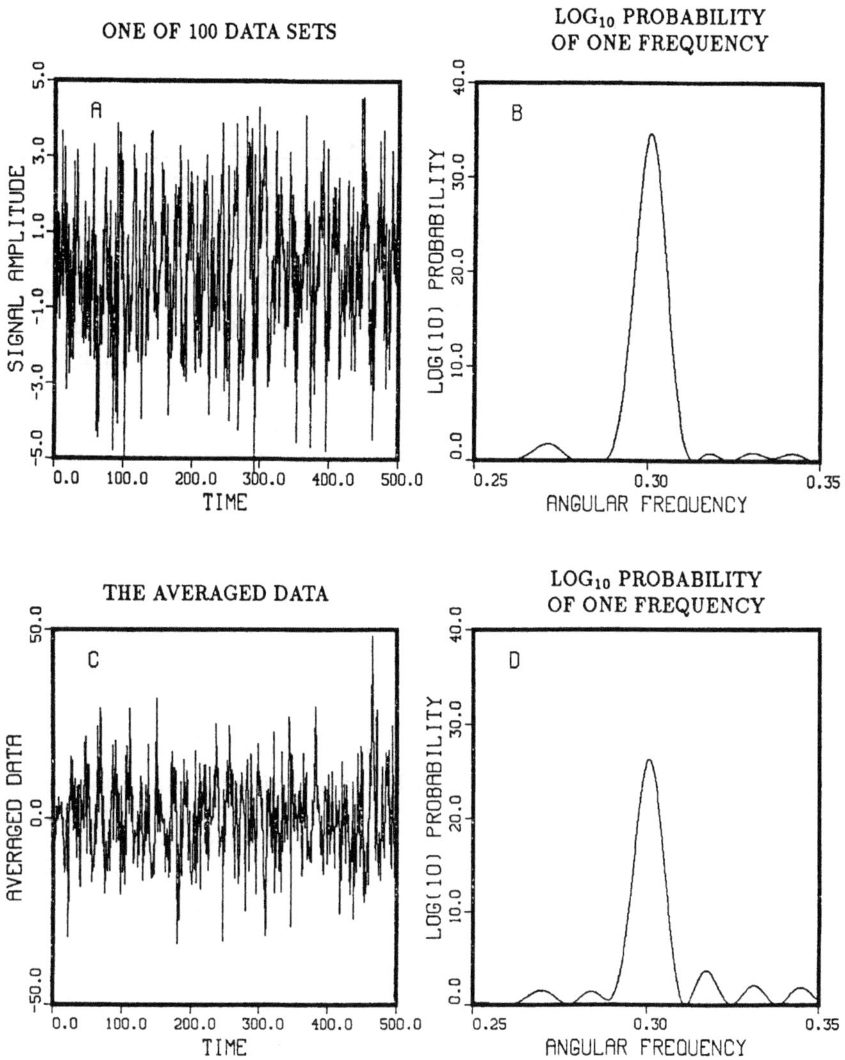

We generated 100 such data sets with different amplitude, phase, and noise variance, but the same frequency (see text for details). In (A) we have displayed one such data set. The $\log_{10}$ of the probability of a single harmonic frequency is displayed in (B). The average data (C) and the $\log_{10}$ probability of a single frequency in (D) show 10^{-9} times less evidence for a harmonic frequency than one data set.

Table 7.2: "Brute Stacking" vs. Joint Analysis

	Frequency Estimate Average	$(\langle\omega\rangle - 0.3)^2$	Frequency Estimate Joint Analysis	$(\langle\omega\rangle - 0.3)^2$
1	0.30018	3.52^{-8}	0.2999974	6.81^{-12}
2	0.29863	1.87^{-6}	0.2999896	1.08^{-10}
3	0.29987	1.47^{-8}	0.2999782	4.75^{-10}
4	0.30076	5.79^{-7}	0.2999995	1.64^{-13}
5	0.30044	2.00^{-7}	0.2999881	1.42^{-10}
6	0.29804	3.82^{-6}	0.2999998	2.78^{-14}
7	0.30024	5.77^{-8}	0.2999995	1.64^{-13}
8	0.30047	2.22^{-7}	0.2999985	2.18^{-12}
9	0.29966	1.09^{-7}	0.2999985	2.18^{-12}
10	0.29990	8.30^{-9}	0.3000088	7.90^{-11}
⋮	⋮	⋮	⋮	⋮
2999	0.30152	2.31^{-6}	0.2999969	9.16^{-12}
3000	0.29968	1.02^{-7}	0.3000008	7.16^{-13}
Average	0.29999	1.39^{-6}	0.2999995	3.02^{-11}
SD		1.17^{-3}		5.49^{-6}

We generated 3000 frequency estimates (see text for details). The frequency estimate in the second column was from the averaged data, and the third column is the variance for that estimate. The fourth and fifth columns are the estimates from the joint analysis. The row labeled "Average" is the average of the 3000 frequency and variance estimates. The last row is the square root of the average variance. Averaging actually appears a little better than it is in these data: when we estimated the frequency we had to supply an initial frequency estimate; this locked the search routine onto the correct peak in the periodogram, even though there was no clear peak above the noise in many of the averaged data sets. From a probability standpoint this is analogous to estimating the average frequency with a strong prior.

at its theoretical best we would expect that in 100 data sets the estimates should improve to $\pm 0.0006/\sqrt{100} = \pm 0.00006$ radians per step. Averaging is a factor of 20 times worse than it should be. But how has the joint analysis done? For 3000 such estimates (in unit signal-to-noise) the joint analysis can do no better than the $\sqrt{n}$ rule: $\pm 0.00006/\sqrt{3000} \approx \pm 0.000001$ radians per step, where we find ± 0.000005, about a factor of 5 larger; we conclude that the mean square weighted amplitude was about 1/25. The joint analysis has performed well; about $0.001/0.000005 = 185$ times better than averaging.

From the 3000 frequency estimates we computed a cumulative distribution of the number of estimates within one standard deviation, two deviations etc. of the true This distribution is displayed in Fig. 7.24. The solid line is the cumulative sample distribution, while the dashed line is the equivalent plot for a Gaussian having the same mean and standard deviation as the sample. The estimates resemble a Gaussian, but there are systematic differences. These differences are numerical in origin. We had to locate the maximum of the posterior probability, and this distribution is roughly 100 times more sharply peaked than a discrete Fourier transform. The pattern search routine we used moves the frequency by some predefined fixed amount. Typically it will move the frequency by only one or two steps, this tends to bunch the estimates up into discrete categories. We could fix this problem at the cost of much greater computing time.

7.5.5 Example – Multiple Measurements

We started this section by presenting a simple diffraction experiment and became sidetracked by some of the implications of the example. When we computed the sufficient statistic of the joint analysis we found the peak to be some 55,000 orders of magnitude higher than the peak for the averaged data. We have used the estimate of the frequency from that peak in several places; here, we plot the results of that analysis to give a better understanding of the determination of the frequency. We will estimate the frequency from the periodogram of the averaged data, from the "Student t-distribution" using the averaged data, and last using a joint analysis on all of the data.

The results of this analysis are displayed in Fig. 7.25. The normalization on all of these curves is arbitrary. If we took the periodogram of the averaged data as our frequency estimate we would have the broad peak in Fig. 7.25. However, probability

Figure 7.24: The Distribution of Sample Estimates

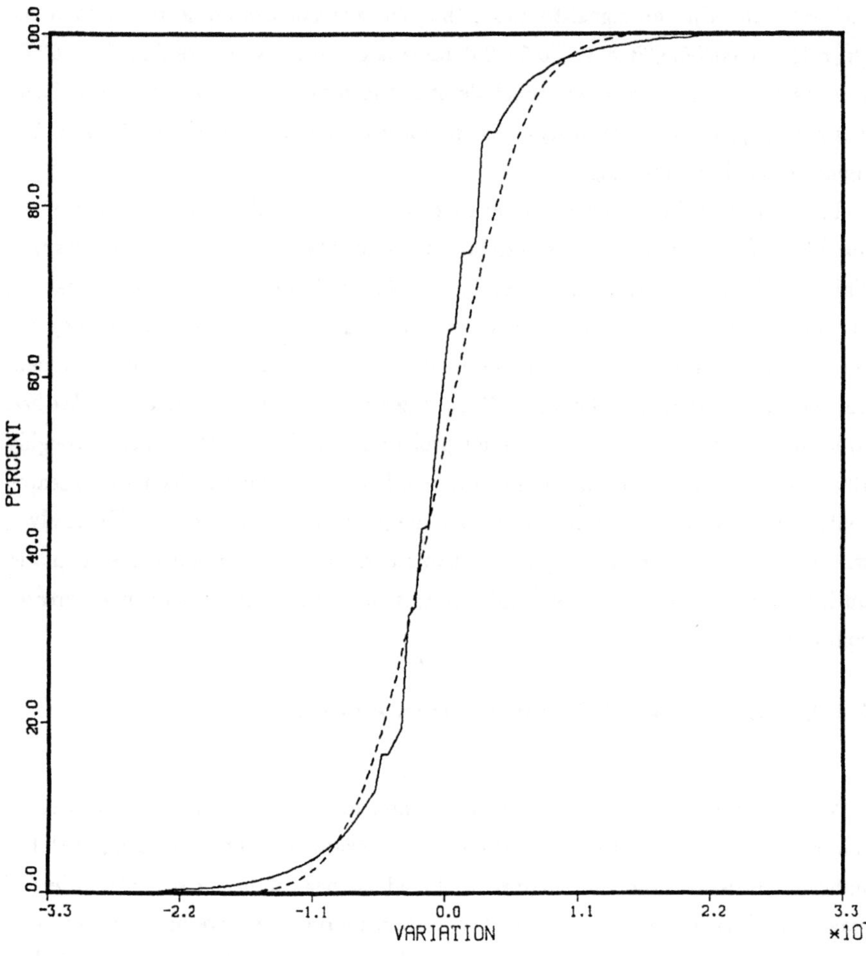

We generated 100 data sets with different amplitude, phase, and noise variance but the same frequency. From the 100 data sets we estimated the frequency. We then generated another 100 data sets and estimated the frequency. We repeated this process some 3000 times. Here we have plotted the cumulative percentage of estimates (solid line) falling within one, two, and three RMS standard deviations. The dashed line is the equivalent distribution for a Gaussian. The axis labels here correspond to two, four, and six standard deviations.

theory applied to the averaged data would narrow that peak by another factor of 10. The resulting posterior distribution is displayed as a sharp Gaussian inside the periodogram. We then estimated the frequency from all of the data using a joint analysis on all 380 data sets. The resulting posterior distribution is displayed as a Gaussian centered at the estimated frequency and having the same variance as our estimate. This is what appears as the vertical line just to the right of the Gaussian from the averaged data. From this we see that the joint analysis estimates the frequency much more precisely than does the analysis of the averaged data, and it estimates it to be rather different from that of the averaged data.

Before leaving this example we would like to apply one more simple analysis to these data. It is true that the small peaks to either side of the main peak in Fig. 7.22 are indications of frequencies. But could there be more that one frequency in the main peak? The spectrum of mercury has many lines in this main peak. Can we see them in these data? To determine whether the main peak has indications for more than one frequency, we compute the probability of two frequencies in this region using only the first five rows from the CCD (we used only the first five rows to reduce computation time). We plotted this as the contour plot in Fig. 7.26. If there is only one frequency present, we expect there to be two ridges in this plot, one extending horizontally and one vertically. On the other hand if there is more than one frequency in these data, there will be a small peak just to one side of $\omega_1 \approx \omega_2$. We see from Fig. 7.26 that there is indeed strong evidence of two frequencies in the main peak. This reinforces one of the things we noted earlier: What one can learn from a data set depends critically on what question one asks. R. A. Fisher once said "let the data speak for themselves". It appears that the data are more than capable of this, but they do not speak spontaneously; they need someone who is willing to ask the right questions, suggested by cogent prior information.

Figure 7.25: Example - Diffraction Experiment

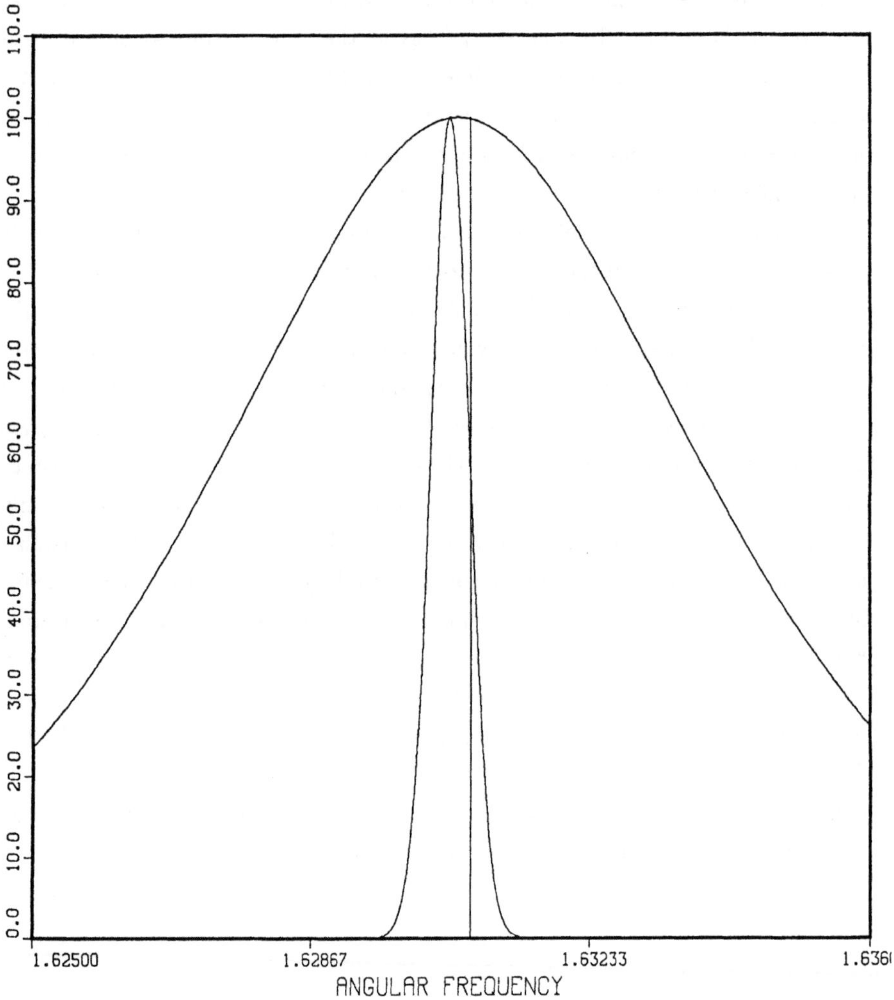

The broad curve on this graph is the periodogram from the averaged data. The sharp Gaussian inside this line is the posterior distribution obtained from the averaged data. The sharp spike located just off-center is the Gaussian representing the posterior distribution from all 380 data sets.

Figure 7.26: Example - Two Frequencies

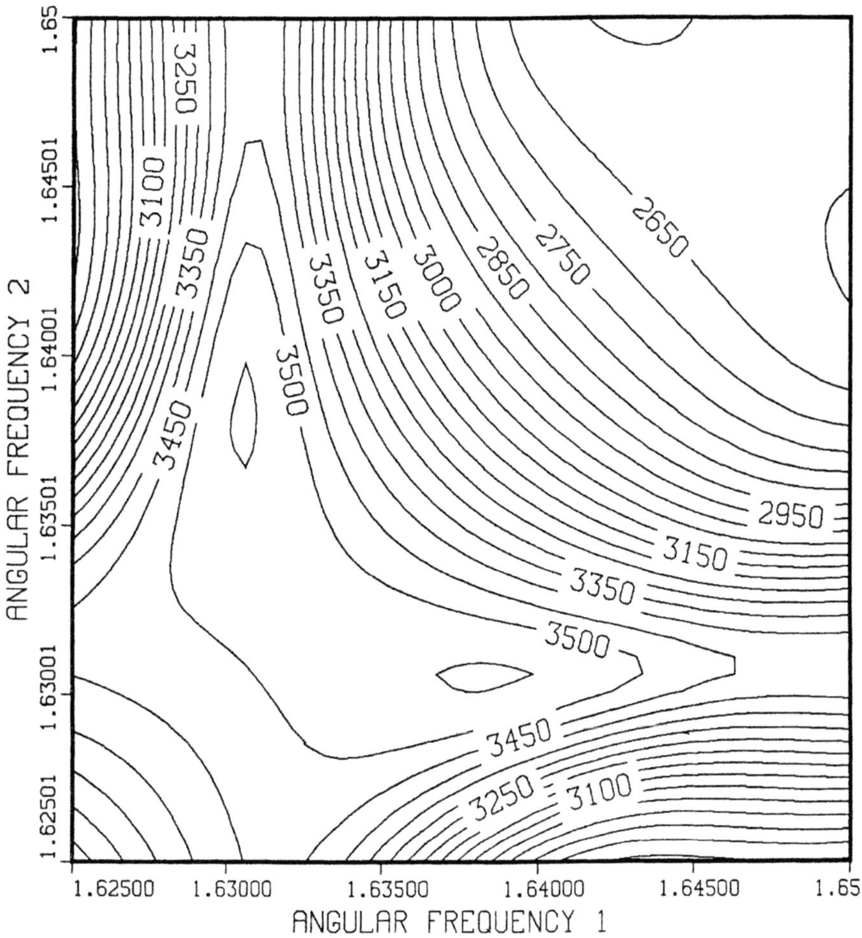

We examined the main peak in the joint analysis to see if there is any evidence for multiple frequencies. The results are presented as a contour plot of the $\log_{10}$ probability of two frequencies in these data. The plot shows clear evidence for two frequencies, even when only a few of the 380 data sets are analyzed, as was done here.

Chapter 8

SUMMARY AND CONCLUSIONS

In this study we have attempted to develop and apply some of the aspects of Bayesian parameter estimation to time series, even though the analysis as formulated is applicable to any data set, be it a time series or not.

8.1 Summary

We began this analysis in Chapter 2, by applying probability theory to estimate the spectrum of a data set that, we postulated, contained only a single sinusoid plus noise. In Chapter 3, we generalized these simple considerations to relatively complex models including the problem of estimating the spectrum of multiple nonstationary harmonic frequencies in the presence of noise. This led us to the "Student t-distribution": the posterior probability of the $\{\omega\}$ parameters, whatever their meaning. In Chapter 4, we estimated the parameters and calculated, among other things, the power spectral density, and the noise variance, and we derived a procedure for assessing the accuracy of the $\{\omega\}$ parameter estimates. In Chapter 6, we specialized to spectrum analysis and explored some of the implications of the "Student t-distribution" for this problem. In Chapter 7, we applied these analyses to a number of real time series with the aim of exploring and broadening some of the techniques needed to apply these procedures. In particular, we demonstrated how to use them to estimate multiple nonstationary frequencies and how to incorporate incomplete information into the estimation problem.

8.2 Conclusions

Perhaps the single biggest conclusion of this work is that what one can learn about a data set depends critically on what questions one asks. If one insists on taking the discrete Fourier transform of a data set, then our analysis shows that one will always obtain good answers to the question "What is the evidence of a single stationary harmonic frequency in these data?" This will be adequate if there are plenty of data and there is no evidence of complex phenomena. However, if the data show evidence for multiple frequencies or complex behavior, the discrete Fourier transform can give misleading or incorrect results in the light of more realistic models.

Although the use of integration to remove nuisance parameters is not new, and indeed the calculation in Chapter 3 has, to some degree, been done by every Bayesian who ever removed a nuisance parameter by integration, the realization of the degree of narrowing of the marginal joint posterior probability density that can be achieved by this is, to the best of our knowledge, new and almost startling. It indicates that, even though we might not be able to estimate an amplitude very precisely, the $\{\omega\}$ parameters often associated with an amplitude may be very precisely estimated. We can often improve the estimation of frequencies and decay rates by orders of magnitude over the estimates obtained from the discrete Fourier transform, least squares, or maximum likelihood. This is not to say that the actual estimates will be very different from those obtained from maximum likelihood or least squares – indeed, when little prior information is available the estimates of the parameters are the maximum likelihood estimates. The major difference is in the indicated accuracy of the estimates.

The principles of least squares or maximum likelihood provide no way to eliminate nuisance parameters, and thus oblige one to seek a global maximum in a space of much high dimensionality, which typically requires orders of magnitude more computation time. Having found this, they provide no way to assess the accuracy of the estimates other than the sampling distribution of the estimator – which is another even longer calculation. But it is a calculation that does not answer the real question of interest; it answers the "pre-data" question:

(Q1): "Before you have seen the data, how much do you expect the estimate to deviate from the true parameter value?"

The question of interest is the "post-data" one:

(Q2): "After getting the data, how accurately does the data set that you actually have determine the true values of the parameters?"

That these are very different questions with different answers in general, was recognized already by R. A. Fisher in the 1930's; he noted that in general two data sets that yield the same numerical value of the estimator, may nevertheless justify very different claims of accuracy. He sought to correct this by his device of conditioning on "ancillary statistics." But Jaynes [42] then showed that this conditioning is mathematically equivalent to using Bayes' theorem, as we have done here. Bayes' theorem, of course, always answers question (Q2), whether or not ancillary statistics exist.

The procedures for comparing models, Eq. (5.9), are perhaps new in the sense that we have extended the Bayesian calculation into the nonlinear $\{\omega\}$ parameters and by carefully keeping track of the normalization constants we were able eventually to integrate out all the parameters. This gives an objective way to compare models and to determine when additional effects are present in the data. Of course, as with any calculation, it will never replace the good sound judgment of the experimenter. The calculation can give a relative ranking of the various choices presented to it. It cannot decide which models to test.

Last, the improvement realized by these procedures when multiple measurements are present is quite striking. The analysis presented in Chapter 7 indicates that the traditional averaging rule will hold whenever the signal is exactly the same in every measurement. Yet in real experiments it is almost impossible to realize the true theoretical improvement. However, by computing the joint marginal posterior probability density of the common effects, the expected $\sqrt{n}$ can be obtained even in data sets where averaging clearly will not work. The implications of this for NMR and other fields are rather profound. Using these techniques we were able to improve resolution in NMR experiments by several orders of magnitude over the discrete Fourier transform; this is making it possible to examine extremely small effects that could not be examined before.

Appendix A

Choosing a Prior Probability

The question "How to choose the prior probability to express complete ignorance?" is interesting in itself, and it cannot be evaded in any problem of scientific inference that is to be solved by using probability theory and Bayes' theorem, but in which we do not wish to incorporate any particular prior information. In the case of the simple harmonic analysis performed in Chapter 2, there are four parameters to be estimated (B_1, B_2, ω, σ), and it is not obvious which choice of prior probabilities is to be preferred. Presumably, any prior probability distribution represents a conceivable state of prior information, but the problem of relating the distribution to the information is subtle and open-ended. You can always think more deeply and thus dredge up more prior information that you didn't think to use at first.

There are two questions one may consider to help in this. First, one should ask "Are the parameters logically connected?" That is, if we gain additional information about one of the parameters, does it change the estimates we would make about the others? If the answer is yes, then the parameters are not logically independent. It will be useful to find a representation where the parameters are independent.

Another useful question is "What are the invariances that the prior probability must obey?" That is, what transformations would convert the present problem into one where we have the same state of prior knowledge? Actually it is only this second question that is truly essential. However, using a representation in which the parameters are not logically independent will mean that the prior probabilities for all the parameters must be determined at once, by utilizing the properties of all the parameters.

In the two representations considered in Chapter 2, Cartesian versus polar, obtaining information about the frequency would rarely affect one's prior estimates of the phase, amplitude, and noise level. Then the prior for the frequency will be independent of the other parameters, and the only invariance to be considered is some group of mappings S of ω onto itself. Later in this appendix we will derive the prior from the group of scale changes.

In the Cartesian representation, B_1 and B_2 are usually logically independent in the sense just noted, so we would assign them independent priors. In the polar notation the amplitude and phase are also logically independent, because obtaining information about either would not affect our prior estimate of the other. The volume elements transform as

$$dB_1 dB_2 = B dB d\theta$$

and so we want a probability density ρ with the two seemingly different forms:

$$\rho(B_1, B_2) dB_1 dB_2 = \rho(B, \theta) B dB d\theta$$

with

$$\rho(B_1, B_2) = f(B_1) f(B_2)$$

but also

$$\rho(B, \theta) = g(B) h(\theta).$$

But we rarely have prior information about θ, so we should take $h(\theta) = \text{const} = 1/2\pi$, $(0 \le \theta \le 2\pi)$. We are left with

$$f(B_1) f(B_2) = \frac{1}{2\pi} g(\sqrt{B_1^2 + B_2^2})$$

but setting $B_2 = 0$, this reduces to

$$f(B_1) f(0) = \frac{1}{2\pi} g(B_1)$$

so we have the functional equation

$$f(x) f(y) = f(\sqrt{x^2 + y^2}) f(0)$$

which a reasonable prior must satisfy. By writing this as

$$\log[f(x)] + \log[f(y)] = \log[f(\sqrt{x^2 + y^2})] + \log[f(0)]$$

the general solution is obvious; if a function $l(x)$ plus a function $l(y)$ is a constant plus a function only of $(x^2 + y^2)$ for all x, y the only possibility is

$$l(x) = ax^2 + b.$$

Thus, $f(x)$ must be a Gaussian; with $a = -1/2\sigma^2$ (the value of b is determined by normalization):

$$f(x) = \frac{1}{\sqrt{2\pi\sigma^2}} \exp\left\{-\frac{x^2}{2\sigma^2}\right\}.$$

To a modern physicist, this argument seems very familiar; it is just a two-dimensional version of Maxwell's original derivation of the Maxwellian velocity distribution [43]. However, historical research has shown that the argument was not original with Maxwell; ten years earlier the astronomer John Herschel [44] had given just our two-dimensional argument in finding the distribution of errors in measuring the position of a star. Thus the Gaussian prior that we use in Appendix B to illustrate the limit as $\sigma \to \infty$ to a uniform prior was not arbitrary; it is the only prior that could have represented our "uninformative" state of knowledge about these parameters. This is a good example of how one can relate prior probabilities to prior information by logical analysis.

In the calculation done by Jaynes [12] the prior used was $dAd\theta$, whereas ours amounts to taking instead $AdAd\theta$. The calculation performed in Chapter 2, and that done by Jaynes will differ from each other only in the fine details. We, effectively, assume slightly less information about the amplitude than Jaynes did, and so we make a slightly less conservative estimate of the frequency. This also simplifies the results by eliminating the Bessel functions found by Jaynes. However, as demonstrated in Appendix B, the differences introduced by the use of different priors to represent ignorance are negligibly small if we have any reasonable amount of data.

When we know that the parameters B_1, B_2, ω, σ are logically independent, how does one choose a prior to represent ignorance of ω and σ? Perhaps the easiest way is to exploit the invariances in the problem. The invariances we would like to exploit are the time invariances. There are two of these: first, the actual starting time of the experiment cannot make any difference; second, a small change in the sampling rate of the problem cannot make any difference provided the same amount of data is collected. To exploit these we apply a technique described by Jaynes [45].

Consider the following problem: we have two experimenters who are to take data on a stationary time series (the same problem described in Chapter 2). Each of these experimenters is free to set up and take the data in any way he sees fit. They do however measure the same time series, starting at slightly different times and using slightly different sampling rates. Now the first experimenter, called E, assigns to his parameters a prior probability

$$P(B_1, B_2, \omega, \sigma|I) \propto G(B_1, B_2, \omega, \sigma) dB_1 dB_2 d\omega d\sigma$$

and the second experimenter called E' assigns to his a prior probability

$$P(B_1', B_2', \omega', \sigma'|I) \propto H(B_1', B_2', \omega', \sigma') dB_1' dB_2' d\omega' d\sigma'.$$

The model equation used by E is just the model used in Chapter 2,

$$f(t, B_1, B_2, \omega) = B_1 \cos(\omega t) + B_2 \sin(\omega t)$$

and E' uses the same equation but with the primed variables

$$f(t', B_1', B_2', \omega') = B_1' \cos(\omega' t') + B_2' \sin(\omega' t').$$

These two equations are related to each other by a simple transformation in the time variable $t' = \alpha t + t_0$ where α is related to the sampling rates and t_0 is the difference in their starting times. The relations between these two system are

$$\begin{gathered} \alpha\omega' = \omega, \qquad \text{and} \qquad \alpha d\omega' = d\omega \\ B_1 = B_1' \cos(\omega' t_0) + B_2' \sin(\omega' t_0) \\ B_2 = B_2' \cos(\omega' t_0) - B_1' \sin(\omega' t_0) \\ dB_1 dB_2 = dB_1' dB_2' \\ \sigma = \gamma\sigma' \qquad \text{and} \qquad d\sigma = \gamma d\sigma'. \end{gathered} \tag{A.1}$$

The factor of α from the time transformation will be absorbed into the frequency as a scaling, because the number of cycles in a given interval $(\omega t/2\pi = \omega' t'/2\pi)$ is an invariant. The squared magnitudes of their model functions are equal; the transformation introduces only an apparent phase change into the signal. In addition to the transformation for the frequency ω the variable σ will have an arbitrary scaling introduced into it.

Now we know that each of these experimenters has performed essentially the same experiment and we expect them to obtain nearly identical conclusions. Each of the experimenters is in the same state of knowledge about his experiment and we apply Jaynes' desideratum of consistency: "In two problems where we have the same prior information, we should assign the same prior probability" [45]. Because E and E' are in the same state of knowledge, H and G are the same functions. Thus we have

$$G(B_1, B_2, \omega, \sigma) dB_1 dB_2 d\omega d\sigma = G(B_1', B_2', \omega', \sigma') dB_1' dB_2' d\omega' d\sigma'.$$

We will solve for the dependence of the prior on the frequency and variance having already obtained the priors for B_1 and B_2. We substitute for ω and σ to obtain

$$G(B_1, B_2, \alpha\omega', \gamma\sigma') = \frac{G(B_1', B_2', \omega', \sigma')}{\gamma\alpha}.$$

This is a functional equation for the prior probability G. It is evident from (A.1) that G must be independent of B_1 and B_2, so the dependence of the prior on the parameters is now completely determined: the only prior which represents complete ignorance of ω, σ, B_1, and B_2 is

$$P(B_1, B_2, \omega, \sigma|I) \propto \frac{1}{\omega\sigma}.$$

This is the Jeffreys prior which we used for the standard deviation σ. Other more cogent derivations of the Jeffreys prior are known [46] but they involve additional technical tools beyond our present scope.

Of course, the realistic limits of the Jeffreys prior do not go all the way to zero and infinity; for example, we always know in advance that σ cannot be less than a value determined by the digitizing accuracy with which we record data; nor so great that the noise power would melt the apparatus. Likewise, as discussed earlier, we know that when the data have zero mean, our data do not contain a zero frequency component; nor can the data contain frequencies so high that they would not pass through our circuitry. Strictly speaking, then, a Jeffreys prior should always be taken between finite positive limits, and be normalized:

$$P(\sigma|I) = \left\{ \begin{array}{ll} A\sigma^{-1} & a < \sigma < b \\ 0 & \text{otherwise} \end{array} \right\} \tag{A.2}$$

with $A^{-1} = \log(b/a)$. But then this prior gets multiplied by a likelihood of the form

$$L(\sigma) = \sigma^{-N} \exp\left\{-\frac{C}{\sigma^2}\right\}$$

which cuts off so strongly as $\sigma \to 0$ and $\sigma \to \infty$ that practically all the mass of the posterior distribution

$$P(\sigma|DI) \propto L(\sigma)P(\sigma|I) \tag{A.3}$$

is concentrated near the peak of (A.3), at $\sigma^2 = 2C/(N+1)$. In our examples, the exact conclusions from (A.2) differ from the limiting ones ($a \to 0, b \to \infty$) by amounts generally less than one part in 10^{20}, so in practice we need never introduce the limits a, b. Similarly, the prior limits on ω have negligible numerical effect, and need not be introduced at all. In our calculation we used a uniform prior for the frequency instead of the Jeffreys prior simply to save writing, because we knew that the difference in the resulting frequency estimates would be negligibly small compared to the width of the posterior distributions (i.e. compared to the error $\delta\omega$ which was inevitable in any event).

Appendix B

Improper Priors as Limits

In the simple harmonic frequency problem Chapter 2 when we removed the amplitudes B_1 and B_2 by integration to get Eq. (2.6), we used a uniform prior probability density which we called an improper prior. In fact, such a function is not a probability density at all. When we use an improper prior, what we really mean is that our prior information is vague, that it carries negligible weight compared to the evidence of the data: the exact prior bounds are so wide that they are far outside the range indicated by the data. To perform the calculation (1.4) correctly, one could bound the parameter to be removed, integrate over the bounded region, and then take a limit as the bound is allowed to go to infinity; but for this problem the result is the same.

Alternatively, we could assume we have a previously measured value of the parameter and then take the limit as the uncertainty in that measurement becomes infinite. We will use a calculation very similar to this in a number of places in the text, and we give this calculation to demonstrate that the use of an improper prior to express "complete ignorance" cannot affect the results in any significant way. This will also show the effect of incorporating additional information into the calculation. Suppose we have some previously measured values for the amplitudes, designated as $\hat{B}_1$ and $\hat{B}_2$. We now proceed to calculate the expectation value of the amplitudes using a prior probability that takes this information into account.

Suppose the previous measured values $\hat{B}_1$ and $\hat{B}_2$ are known with an accuracy of $\pm\delta$ (interpreted as the standard deviation of a Gaussian error distribution for the previous measurements). The joint prior probability density of the true values B_1

and B_2 is the posterior distribution for the first measurement,

$$P(B_1, B_2|I) = \left[2\pi\delta^2\right]^{-1} \exp\left\{-\frac{1}{2\delta^2}[(\hat{B}_1 - B_1)^2 + (\hat{B}_2 - B_2)^2]\right\} \tag{B.1}$$

which becomes our informative prior for the second measurement. Then using Bayes' theorem, the posterior probability of the parameters is proportional to the product of the prior (B.1) and the likelihood (2.3):

$$P(B_1, B_2|D, I) = \left[2\pi\delta^2\right]^{-1} \left[2\pi\sigma^2\right]^{-\frac{N}{2}} \exp\left\{-\frac{X}{2\delta^2} - \frac{NY}{4\sigma^2}\right\}$$

where

$$X \equiv (\hat{B}_1 - B_1)^2 + (\hat{B}_2 - B_2)^2$$

$$Y \equiv B_1^2 + B_2^2 - 2\left(\frac{2R(\omega)}{N}B_1 + \frac{2I(\omega)}{N}B_2\right).$$

After a little algebra the posterior probability may be written as

$$P(B_1, B_2|D, I) = \left[2\pi\delta^2\right]^{-1} \left[2\pi\sigma^2\right]^{-\frac{N}{2}} \exp\left\{-\beta[(B_1 - E_1)^2 + (B_2 - E_2)^2]\right\}$$

$$\beta = \frac{\delta^2 + \sigma^2}{2\delta^2\sigma^2}$$

where

$$E_1 = \frac{\delta^2[2R(\omega)/N] + \sigma^2\hat{B}_1}{\delta^2 + \sigma^2} \tag{B.2}$$

$$E_2 = \frac{\delta^2[2I(\omega)/N] + \sigma^2\hat{B}_2}{\delta^2 + \sigma^2} \tag{B.3}$$

are the posterior expectations:

$$\langle B_1 \rangle = E_1 \qquad \text{and} \qquad \langle B_2 \rangle = E_2.$$

The posterior estimates are now weighted averages of the two measurements. This is a rather old result, first discovered by Laplace [47] but essentially forgotten for a century, until the modern development of Bayesian methods began to demonstrate that most of Laplace's results were correct and important.

To understand the full implications of this we will consider three special cases. First, when $\delta \ll \sigma$, the previous measurement is much better than the current one. Then

$$\langle B_1 \rangle = \hat{B}_1 \qquad \text{and} \qquad \langle B_2 \rangle = \hat{B}_2$$

which says to use the original measured value, a most pleasing result, since that is exactly what any physicist would have done anyway. Second, consider the case where $\sigma = \delta$. Then

$$\langle B_1 \rangle = \frac{1}{2}\left(\frac{2R(\omega)}{N} + \hat{B}_1\right) \qquad \text{and} \qquad \langle B_2 \rangle = \frac{1}{2}\left(\frac{2I(\omega)}{N} + \hat{B}_2\right)$$

which says the two measurements are of equal weight and one should average them. Again a most pleasing result, since that is exactly what one's intuition would have told one to do. Third, consider the case when $\delta \gg \sigma$ (one knows only that the two amplitudes must be bounded) then,

$$\langle B_1 \rangle = \frac{2R(\omega)}{N} \qquad \text{and} \qquad \langle B_2 \rangle = \frac{2I(\omega)}{N}. \tag{B.4}$$

This is the result obtained using the improper prior. In the limit as δ goes to infinity, the prior (B.1) goes smoothly into the uniform improper prior used in our calculation of Eq. (2.6), and the weighted averages go smoothly into (B.4).

The important point here is that if δ is appreciably greater than σ, the prior we use does not make any significant difference; as δ becomes larger, less information is conveyed by the prior measurement, and probability theory as indicated by (B.2) and (B.3) automatically assigns less weight to it. The result must depend mostly on the evidence in the data. In the limit as δ goes to infinity we have incorporated no prior information about the parameter, and the result must depend totally on the data.

Appendix C

Removing Nuisance Parameters

We illustrate in this appendix that integrating over a nuisance parameter is very much like estimating the parameter from the data and constraining it in the posterior probability to that value. We first estimate the amplitudes by calculating their posterior expectations, and then substitute them into the likelihood (2.3). If integrating over a nuisance parameter is nearly the same, we should obtain (2.6), or at the very least something very much like (2.6). We assume for this illustration that σ is known; then, using the likelihood Eq. (2.3), the expectation value of B_j, supposing ω known, is

$$\langle B_j \rangle = \frac{\int_{-\infty}^{+\infty} dB_1 dB_2 B_j L(B_1, B_2, \omega, \sigma)}{\int_{-\infty}^{+\infty} dB_1 dB_2 L(B_1, B_2, \omega, \sigma)}. \tag{C.1}$$

We take these as our estimates $\langle B_j \rangle$ (ω) in (2.3). Carrying out the required integrations gives the posterior expectation values of B_1 and B_2:

$$B_1^*(\omega) = \langle B_1(\omega) \rangle = \frac{2R(\omega)}{N}, \tag{C.2}$$

$$B_2^*(\omega) = \langle B_2(\omega) \rangle = \frac{2I(\omega)}{N},$$

where $R(\omega)$ and $I(\omega)$ are the cosine and sine transforms of the data, as defined in (2.4) and (2.5). Now these are substituted back into (2.3) to give

$$L(B_1^*, B_2^*, \omega, \sigma) \propto \sigma^{-N} \exp\left\{-\frac{N}{2\sigma^2}[\overline{d^2} - \frac{2}{N}C(\omega)]\right\}. \tag{C.3}$$

But in its dependence on ω, this is just (2.6): integrating over the amplitudes with respect to the uniform prior has given us the same result as constraining them to their expectation values (C.1).

The two procedures are not always equivalent, as they happen to be here, but they can never be very different whenever we have enough information or data to make a good estimate of a nuisance parameter. In fact, these procedures would have been slightly different in this example if we had not assumed the noise variance σ^2 to be known. Then σ^2 would also become a nuisance parameter which we would remove by integration, and the "Student t-distribution" thus obtained would be raised to the $-N/2$ power instead of $(2-N)/2$ as was found in Chapter 2.

More generally, whenever a nuisance parameter is actually well determined by many data ($N \to \infty$), these two procedures become for all practical purposes equivalent. But when the data are too meager to determine the nuisance parameters very well, the *ad hoc* procedure (C.3) can be overoptimistic, leading us to think that we have determined ω more accurately than the data really justify; and if we have relevant prior information about the parameters the *ad hoc* method ignores it.

Appendix D

Uninformative Prior Probabilities

When we worked the single frequency problem in Chapter 2 we used a uniform prior for the amplitudes. In polar coordinates this prior is

$$P(B, \theta | I) \propto B dB d\theta$$

and leads to

$$P(\omega | \sigma, D, I) \propto \exp\left\{\frac{C(\omega)}{\sigma^2}\right\} \tag{D.1}$$

as the posterior probability of a single harmonic frequency, given the data and the noise variance σ^2. When Jaynes [12] worked this problem he performed the calculation in polar coordinates and supposed prior information I' for which

$$P(B\theta | I') \propto dB d\theta$$

as the prior for the amplitude and phase. He then arrived at

$$P(\omega | \sigma, D, I') \propto \exp\left\{\frac{C(\omega)}{2\sigma^2}\right\} I_0\left(\frac{C(\omega)}{2\sigma^2}\right) \tag{D.2}$$

where I_0 is a Bessel function of order zero. This is a very different looking result, given that the only difference in the two calculations was the prior used. How can such a simple change in the problem have such a dramatic effect on the answer, and just what effect did the use of these two different priors have on the results?

The main question we will pursue here is "What effect did this different prior have on the frequency estimate?" The answer to this question is surprising: since Eq. (D.1) and Eq. (D.2) are both functions of $C(\omega)$, they both reach their maximum at the same value $\omega = \hat{\omega}$; there is no difference at all in the actual frequency estimate! But there is a difference in the curvatures of Eq. (D.1) and Eq. (D.2) at their common maximum $\hat{\omega}$, so there is a difference in the claimed accuracy of that estimate. Recalling that in the Gaussian approximation it is the second derivative of $\log(P(\omega|\sigma, D, I)$ that matters,

$$\frac{d^2}{d\omega^2}\log P(\omega|\sigma, D, I)\bigg|_{\omega=\hat{\omega}} = \frac{1}{(\delta\omega)^2}$$

a short calculation gives for the standard deviations, from Eq. (D.1)

$$\delta\omega = \frac{\sigma}{\sqrt{C''(\hat{\omega})}}$$

and from Eq. (D.2)

$$\delta\omega' = \frac{\sigma}{\sqrt{C''(\hat{\omega})}}\left(\frac{2I_0}{I_0 + I_1}\right)^{\frac{1}{2}}$$

where the argument of the I_0 and I_1 Bessel functions is $C(\hat{\omega})/2\sigma^2$. The ratio of the error estimates is $q(C(\hat{\omega}/2\sigma^2))$, where

$$q(x) = \left(\frac{2I_0(x)}{I_0(x) + I_1(x)}\right)^{\frac{1}{2}}.$$

Substituting some numerical values for x we have

x	$q(x)$
0	1.414
1	1.176
2	1.086
4	1.036
8	1.016
> 18	$1 + (8x)^{-1}$.

Now if there is a single sinusoid present with amplitude B, the maximum of the periodogram will be about

$$C(\hat{\omega}) \approx \frac{NB^2}{4}.$$

With a signal-to-noise ratio of unity, the mean square signal $B^2/2 = \sigma^2$, so

$$\frac{C(\hat{\omega})}{2\sigma^2} \approx \frac{N}{4}.$$

If $N \geq 10$, there is less than a 6.5% difference in the error estimates, and when $N > 50$ the difference is less than 1%. Thus whenever we have enough signal-to-noise ratio or enough data to justify any frequency estimates at all, the differences are completely negligible.

Appendix E

Computing the "Student t-Distribution"

This subroutine was used to prepare all of the numerical analysis presented in this work. This is a general purpose implementation of the calculation that will work for any model functions and for any setting of the parameters, independent of the number of parameters and their values, and it does not care if the data are uniformly sampled or not. In order to do this, the subroutine requires five pieces of input data and one work area. On return one receives $H_i(t_j)$, h_i, $\overline{h^2}$, $P(\{\omega\}|D,I)$, $\langle\sigma\rangle$, and $\hat{p}(\{\omega\})$. The parameter list is as follows:

Parm	LABEL	i/o	Description/function
N	INO	input	The number of discrete time samples in the time series to be analyzed.
m	IFUN	input	This is the order of the matrix g_{jk} and is equal to the number of model functions.
d_j	DATA	input	The time series (length N): this is the data to be analyzed. Note: the routine does not care if the data are sampled uniformly or not.
g_{ij}	GIJ	input	This matrix contains the j nonorthogonal model functions [dimensioned as GIJ(INO,IFUN)] and evaluated at t_i.

Parm	LABEL	i/o	Description/function
ZLOGE	ZLOGE	i/o	This is the $\log_e$ of the normalization constant. The subroutine never computes the "Student t-distribution" when ZLOGE is zero: instead the $\log_e$ of the "Student t-distribution" is computed. It is up to the user to locate a value of $\log_e[P(\{\omega\}\|D,I)]$ close to the maximum of the probability density. This log value should then be placed in ZLOGE to act as an upper bound on the normalization constant. With this value in place the subroutine will return the value of the probability; then, an integral over the probability density can be done to find the correct value of the normalization constant.
$H_i(t_j)$	HIJ	output	These are orthonormal model functions Eq. (3.5) evaluated at the same time and parameter values as GIJ.
h_i	HI	output	These are projections of the data onto the orthonormal model functions Eq. (3.13) and Eq. (4.3).
$\overline{h^2}$	H2BAR	output	The sufficient statistic $\overline{h^2}$ Eq. (3.15) is always computed.
$P(\{\omega\}\|D,I)$	ST	output	The "Student t-distribution" Eq. (3.17) is not computed when the normalization constant is zero. To insure this field is computed the normalization constant must be set to an appropriate value.
STLE	STLE	output	This is the $\log_e$ of the "Student t-distribution" Eq. (3.17) and is always computed.
$\langle\sigma\rangle$	SIG	output	This is the expected value of the noise variance σ as a function of the $\{\omega\}$ parameters Eq. (4.6) with $s = 1$.
$\hat{p}(\{\omega\})$	PHAT	output	This is the power spectral density Eq. (4.15) as a function of the $\{\omega\}$ parameters.

Parm	LABEL	i/o	Description/function
	WORK	scratch	This work area must be dimensioned at least $5m^2$. The dimension in the subroutines was set high to avoid possible "call by value" problems in FORTRAN. On return, WORK contains the eigenvectors and eigenvalues of the g_{jk} matrix. The eigenvector matrix occupies m^2 contiguous storage locations. The m eigenvalues immediately follow the eigenvectors.

This subroutine makes use of a general purpose "canned" eigenvalue and eigenvector routine which has not been included. The original routine used was from the IMSL library and the code was later modified to use a public-domain implementation (an EISPACK routine). The actual routine one uses here is not important so long as the routine calculates both the eigenvalues and eigenvectors of a real symmetric matrix. If one chooses to implement this program one must replace the call (clearly marked in the code) with a call to an equivalent routine. Both the eigenvalues and eigenvectors are used by the subroutine and it assumes that the eigenvectors are normalized.

```
      SUBROUTINE PROB
     C (INO,IFUN,DATA,GIJ,ZLOGE,HIJ,HI,H2BAR,ST,STLOGE,SIGMA,PHAT,WORK)
      IMPLICIT REAL*08(A-H,O-Z)
      DIMENSION DATA(INO),HIJ(INO,IFUN),HI(IFUN),GIJ(INO,IFUN)
      DIMENSION WORK(IFUN,IFUN,20)
C
C
      CALL VECTOR(INO,IFUN,GIJ,HIJ,WORK)
C
      H2=0D0
      DO 1600 J=1,IFUN
      H1=0D0
      DO 1500 L=1,INO
 1500 H1=H1 + DATA(L)*HIJ(L,J)
      HI(J)=H1
      H2=H2 + H1*H1
 1600 CONTINUE
      H2BAR=H2/IFUN
C
      Y2=0D0
      DO 1000 I=1,INO
 1000 Y2=Y2 + DATA(I)*DATA(I)
      Y2=Y2/INO
```

```
C
      QQ=1D0 - IFUN*H2BAR / INO / Y2
      STLOGE=DLOG(QQ) * ((IFUN - INO)/2D0)
C
      AHOLD=STLOGE - ZLOGE
      ST   =0D0
      IF(DABS(ZLOGE).NE.0D0)ST=DEXP(AHOLD)
C
      SIGMA=DSQRT( INO/(INO-IFUN-2) * (Y2 - IFUN*H2BAR/INO) )
C
      PHAT = IFUN*H2BAR * ST
C
      RETURN
      END
      SUBROUTINE VECTOR(INO,IFUN,GIJ,HIJ,WORK)
      IMPLICIT REAL*8(A-H,O-Z)
      DIMENSION HIJ(INO,IFUN),GIJ(INO,IFUN),WORK(IFUN,IFUN,20)
C
      DO 1000 I=1,IFUN
      DO 1000 J=1,INO
 1000 HIJ(J,I)=GIJ(J,I)
C
      CALL ORTHO(INO,IFUN,HIJ,WORK)
C
      DO 5000 I=1,IFUN
      TOTAL=0D0
      DO 4500 J=1,INO
 4500 TOTAL=TOTAL + HIJ(J,I)**2
      ANORM=DSQRT(TOTAL)
      DO 4000 J=1,INO
 4000 HIJ(J,I)=HIJ(J,I)/ANORM
 5000 CONTINUE
C
      RETURN
      END

      SUBROUTINE ORTHO(INO,NMAX,AIJ,W)
      IMPLICIT REAL*8 (A-H,O-Z)
      REAL*8  AIJ(INO,NMAX),W(NMAX)
C
      IT=1
      IE=IT + NMAX*NMAX
      IM=IE + NMAX*NMAX
      IW=IM + NMAX*NMAX
      I2=IW + NMAX*NMAX
C
      CALL TRANS(INO,NMAX,AIJ,W(IM),W(IT),W(IE),W(IW),W(I2))
```

```
C
      RETURN
      END
      SUBROUTINE TRANS
     C(INO,NMAX,AIJ,METRIC,TRANSM,EIGV,WORK1,WORK2)
      IMPLICIT REAL*8 (A-H,O-Z)
      REAL*8  AIJ(INO,NMAX)
      REAL*8  METRIC(NMAX,NMAX),EIGV(NMAX)
      REAL*8  TRANSM(NMAX,NMAX),WORK1(NMAX),WORK2(NMAX)
      DO 2000 I=1,NMAX
      DO 2000 J=1,NMAX
      TOTAL=0D0
      DO 1000 K=1,INO
 1000 TOTAL=TOTAL + AIJ(K,I)*AIJ(K,J)
      METRIC(I,J)=TOTAL
 2000 CONTINUE
C******************************************************************
C**** THIS CALL MUST BE REPLACED WITH THE CALL TO AN EIGENVALUE
C**** AND EIGENVECTOR ROUTINE
      CALL EIGERS(NMAX,NMAX,METRIC,EIGV,1,TRANSM,WORK1,WORK2,IERR)
C**** NMAX   IS THE ORDER OF THE MATRIX
C**** METRIC IS THE MATRIX FOR WHICH THE EIGENVALUES AND VECTORS
C****        ARE NEEDED
C**** EIGV   MUST CONTAIN THE EIGENVALUES ON RETURN
C**** TRANSM MUST CONTAIN THE EIGENVECTORS ON RETURN
C**** WORK1  IS A WORK AREA USED BY MY ROUTINE AND MAY BE USED
C****        BY YOUR ROUTINE.  ITS  DIMENSION IS NMAX
C****        IN THIS ROUTINE. HOWEVER IT MAY BE DIMENSIONED
C****        AS LARGE AS NMAX*NMAX WITHOUT AFFECTING ANYTHING.
C**** WORK2  IS A SECOND WORK AREA AND IS OF DIMENSION NMAX
C****        IN THIS ROUTINE, IT MAY ALSO BE DIMENSIONED AS
C****        LARGE AS NMAX*NMAX WITHOUT AFFECTING ANYTHING.
C******************************************************************
      DO 5120 K=1,INO
      DO 3100 J=1,NMAX
 3100 WORK1(J)=AIJ(K,J)
      DO 5120 I=1,NMAX
      TOTAL=0D0
      DO 3512 J=1,NMAX
 3512 TOTAL=TOTAL + TRANSM(J,I)*WORK1(J)
 5120 AIJ(K,I)=TOTAL
      RETURN
      END
```

Bibliography

[1] Bretthorst G. L., (1987), Bayesian Spectrum Analysis and Parameter Estimation, Ph.D. thesis, Washington University, St. Louis, MO., available from University Microfilms Inc., Ann Arbor Mich.

[2] Robinson, E. A., (1982), "A Historical Perspective of Spectrum Estimation," *Proceedings of the IEEE,* 70, pp. 855-906.

[3] Marple, S. L., (1987), Digital Spectral Analysis with Applications, Prentice-Hall, New Jersey.

[4] Laplace, P. S., (1812), Théorie Analytique des Probabilités, Paris, (2nd edition, 1814; 3rd edition, 1820).

[5] Legendre, A. M., (1806), "Nouvelles Méthods pour la Détermination des Orbits des Cométes," Paris.

[6] Gauss, K. F., (1963 reprint) Theory of the Motion of the Heavenly Bodies Moving About the Sun in Conic Sections, Dover Publications, Inc., New York.

[7] Cooley, J. W., P. A. Lewis, and P. D. Welch, (1967), "Historical Notes on the Fast Fourier Transform," *Proc. IEEE* 55, pp. 1675-1677.

[8] Brigham, E., and R. E. Morrow, (1967), "The Fast Fourier Transform," *Proc. IEEE Spectrum,* 4, pp. 63-70.

[9] Gentleman, W. M., (1968), "Matrix Multiplication and Fast Fourier Transformations," *Bell Syst. Tech. Journal,* 17, pp. 1099-1103.

[10] Cooley, J. W., and J. W. Tukey, (1965), "An Algorithm for the Machine Calculation of Complex Fourier Series," *Mathematics of Computation,* 19, pp. 297-301.

[11] Schuster, A., (1905), "The Periodogram and its Optical Analogy," *Proceedings of the Royal Society of London,* 77, pp. 136.

[12] Jaynes, E. T. (1987), "Bayesian Spectrum and Chirp Analysis," inMaximum Entropy and Bayesian Spectral Analysis and Estimation Problems, C. Ray Smith, and G. J. Erickson, ed., D. Reidel, Dordrecht-Holland, pp. 1-37.

[13] Blackman, R. B., and J. W. Tukey, (1959), The Measurement of Power Spectra, Dover Publications, Inc., New York.

[14] Jaynes, E. T. (1983), Papers on Probability, Statistics and Statistical Physics, a reprint collection, D. Reidel, Dordrecht-Holland.

[15] Jeffreys, H., (1939), Theory of Probability, Oxford University Press, London, (Later editions, 1948, 1961).

[16] Lord Rayleigh, (1879), *Philosophical Magazine,* 5, pp. 261.

[17] Tukey, J. W., several conversations with E. T. Jaynes, in the period 1980-1983.

[18] Waldmeier, M., (1961), The Sunspot Activity in the Years 1610-1960, Schulthes, Zurich.

[19] Nyquist, H., (1928), "Certain Topics in Telegraph Transmission Theory," *Transactions AIEE*, pp. 617.

[20] Nyquist, H., (1924), "Certain Factors Affecting Telegraph Speed," *Bell System Technical Journal,* 3, pp. 324.

[21] Hooke, R., and T. A. Jeeves, (1962), "Direct Search Solution of Numerical and Statistical Problems," *J. Assoc. Comp. Mach.,* pp. 212-229.

[22] Wilde D. J., (1964), Optimum Seeking Methods, Prentice-Hall, Inc. Englewood Cliffs, N. J.

[23] Zellner, A., (1980), in Bayesian Statistics, J. M. Bernardo, ed., Valencia University Press, Valencia, Spain.

[24] Geisser, S., and J. Cornfield, (1963), "Posterior Distribution for Multivariate Normal Parameters," *Journal of the Royal Statistical Society*, B25, pp. 368-376.

[25] Zellner, A., (1971), An Introduction to Bayesian Inference in Econometrics, John Wiley and Sons, New York. Second edition, (1987).

[26] Cox, R. T., (1961), The Algebra of Probable Inference, Johns-Hopkins Press, Baltimore, Md.

[27] Tribus, M., (1969), Rational Descriptions, Decisions and Designs, Pergamon Press, Oxford.

[28] Schlaifer, R., (1959), Probability and Statistics for Business Decisions: an Introduction to Managerial Economics Under Uncertainty, McGraw-Hill Book Company, New York.

[29] Whittle, P., (1954), Appendix to H. Wold, Stationary Time Series, Almquist and Wiksell, Stockholm, pp. 200-227.

[30] Shaw, D., (1976), Fourier Transform NMR Spectroscopy, Elsevier Scientific Pub. Co., New York.

[31] Ganem, J. W., and R. E. Norberg, (1987), Private Communication.

[32] Abragam, A., (1961), Principles of Nuclear Magnetism, Oxford Science Publications, London.

[33] Beckett, R. J., (1979), The Temperature and Density Dependence of Nuclear Spin-Spin Interactions in Hydrogen-Deuteride Gas and Fluid, Ph.D. thesis, Rutgers University, New Brunswick, New Jersey; available from University Microfilms Inc., Ann Arbor Mich.

[34] Currie, R. G., (1985), Private Communication.

[35] Currie, R. G., and S. Hameed, (1986), "Climatically Induced Cyclic Variations in United States Corn Yield and Possible Economic Implications," presented at the Canadian Hydrology Symposium, Regina, Saskatchewan.

[36] Burg, John Parker, (1975), Maximum Entropy Spectral Analysis, Ph.D. Thesis, Stanford University; available from University Microfilms Inc., Ann Arbor Mich.

[37] Cohen, T. J., and P. R. Lintz, (1974), "Long Term Periodicities in the Sunspot Cycle," *Nature,* 250, pp. 398.

[38] Sonett, C. P., (1982), "Sunspot Time Series: Spectrum From Square Law Modulation of the Half Cycle," *Geophysical Research Letters,* 9 pp. 1313-1316.

[39] Bracewell, R. N,. (1986), "Simulating the Sunspot Cycle," *Nature,* 323, pp. 516.

[40] Jaynes, E. T., (1982), "On the Rationale of Maximum-Entropy Methods", *Proceedings of the IEEE,* 70, pp. 939-952.

[41] Smith, W. H., and W. Schempp, (1987) private communication.

[42] Jaynes, E. T., (1976), "Confidence Intervals vs. Bayesian Intervals," in Foundations of Probability Theory, Statistical Inference, and Statistical Theories of Science, W. L. Harper and C. A. Hooker, editors, D. Reidel Publishing Co., pp. 252-253; reprinted in [14].

[43] Maxwell, J. C., (1860), "Illustration of the Dynamical Theory of Gases. Part I. On the Motion and Collision of Perfectly Elastic Spheres," *Philosophical Magazine,* 56.

[44] Herschel, J., (1850), *Edinburgh Review,* 92, pp. 14.

[45] Jaynes, E. T., (1968), "Prior Probabilities," *IEEE Transactions on Systems Science and Cybernetics,* SSC-4, pp. 227-241; reprinted in [14].

[46] Jaynes, E. T., (1980), "Marginalization and Prior Probabilities," in *Bayesian Analysis in Econometrics and Statistics,* A. Zellner, ed., North-Holland Publishing Company, Amsterdam; reprinted in [14].

[47] Laplace, P. S., (1814), A Philosophical Essay on Probabilities, Dover Publications, Inc., New York, (1951, unabridged and unaltered reprint of Truscott and Emory translation).

Index

Δt *22*
λ_l *33*
σ *15*, 46, 47
$\overline{\omega^2}$ *61*
$\hat{\omega}$ *50*
$\{\omega\}$ *2*, 48
A_k *33*, 44
Abragam, A. 118, 144
absorption spectrum 132, 134
accuracy estimates 20, 27, 50, 86, 98, 102, 100, 167
aliasing 81
amplitudes
  nonorthonormal 13, *31*
  orthonormal *33*
applications
  chirp analysis 158
  decay envelope extraction 144
  economic 134
  harmonically related frequencies 157
  multiple frequency estimation 151
  multiple measurements 161
  NMR 117, 144
  nonstationary frequency estimation 117
  orthogonal expansion 148
assumptions violating 74
averaging data 163
b *20*
B_j *31*, 44
Bayes theorem *8*, 16, 55, 57
Beckett, R. J. 134
Bessel inequality 35
Blackman, R. B. 9, 23, 73
Blackman-Tukey spectral estimate 72
Bracewell, R. N. 151, 158
Brigham, E. 6
Burg algorithm 135, 151
Burg, J. P. 135, 151
chirp 159
$C(\omega)$ *7*
Cohen, T. J. 148
complete ignorance
  choosing a prior 183
  of a location parameter 18, 185
  of a scale parameter 19, 187
Cooley, J. W. 6
Cornfield, J. 73
cosine transform 7, *16*
Cox, R. T. 76
Currie, R. G. 135
D *9*
d_i *13*, 31
$\overline{d^2}$ *17*
data
  corn 135
  covariances 37
  diffraction 162
  economic 134
  NMR 118, 144
direct probability *9*, 31
discrete Fourier transform *7*, 19, 89, 92, 105, 108, 110
energy 25, 51
expected
  $\{\omega\}$ Parameters 48
  amplitudes nonorthonormal 44
  amplitudes orthogonal 44
  variance 46
$f(t)$ *13*, 31
Fisher, R. A. 74, 175
frequency estimation

common 120
multiple 151
one 13
g_{jk} *32*
Gauss, K. F. 5
Gaussian *15*
Gaussian approximation 20, 49, 88, 98
Geisser, S. 73
Gentleman, W. M. 6
$\overline{h^2}$ *35*
h_j *34*
H *9*
$H_j(t)$ *33*
Hanning window 23, 73
Herschel, J. 185
Hooke, R. 50
hyperparameter 59
I *9*
$I(\omega)$ 7, *16*, 71, 88, 97, 193
improper prior
Jeffreys 19
uniform 18
intuitive picture 80
Jaynes, E. T. 7, 13, 14, 16, 21, 23, 31, 50, 69, 98, 110, 151, 181, 185, 186, 187, 195
Jeffreys prior 19, 35, 46, 187
Jeffreys, H. 19, 55
joint quasi-likelihood 18, 34
Laplace, P. S. 5, 190
least squares 2, 16
Legendre, A. M. 5
Lewis, A. 6
likelihood *9*
general model 31
global 61
one-frequency 13
ratio 64
line power spectral density *114*
Lintz, P. R. 148
location parameter 18, 185
m *31*
marginal posterior probability definition 10
Marple, S. L. 5, 8, 27
maximum entropy 14
maximum likelihood 2, 16
Maxwell, J. C. 185
mean-square
$\{\hat{\omega}\}$ *61*
d_i *17*
h_j *35*
model *13*
adequacy 38
Bracewell's 158
chipped frequency 158
decay envelope 144
harmonically related frequencies 157
intuitive picture 36
multiple harmonic frequencies 108
multiple nonstationary frequencies 115, 120
one-frequency *13*, 70
with a chirp 159
with a constant 137, 151
with a Lorentzian decay 86, 122
with a trend 137
orthonormal *33*
selection 55
two-frequencies 94
Morrow, R. E. 6
multiple frequency 108
multiple measurements 120, 161
noise 15, 78
nonuniform sampling 81, 83
Norberg, R. E. 118, 144
nuisance function 137
nuisance parameter *10*, 18, 34, 146, 193
Nyquist, H. 27
Occam's razor 64
orthnormality 33
orthogonal expansion 64
orthogonal projection 34
orthonormal model
one-frequency 71
one-frequency Lorentzian decay 87
two-frequency 97
$\hat{p}(\{\omega\})$ 25, 51

pattern search routine 50
periodogram *7*, 18, 25, 72, 82, 92, 105, 110, 153
posterior covariances
  $\{\omega\}$ 50
  $\{A\}$ 45
posterior odds ratio 64, 107, 122
posterior probability *9*
  approximate 40, 49
  f_j 57, 61, 63
  general 35
  multiple measurements 135, 167
  multiple well separated frequencies 110
  of one-frequency with Lorentzian decay 87
  of one-frequency 18, 71
  of the expansion order 149
  of two-frequencies 94, 98, 105
  of two-frequencies with trend 138
  of two well separated frequencies 95
power spectral density 25, *51*, 72, 103, 112
prior probability *9*
  ω_j 60
  $\{A\}$ 59
  assigning 14, 183
  complete ignorance 183, 195
  Gaussian 185, 190
  improper priors as limits 189
  incorporation prior information 18, 34, 59, 190
  Jeffreys 35, 187
  uniform 18, 34, 185, 189
prior see prior probability
product rule 9, 120
quadrature data 117
$R(\omega)$ 7, *16*, 71, 97, 88, 193
R_α 62
Rayleigh criterion 23
relative probabilities 56, 63, 93
residuals definition 5
Robinson, E. A. 5
$\hat{S}(\omega)$ *114*
sampling distribution *9*, 37
scale parameter 19, 187
Schempp, W. 162
Schlaifer, R. 76
Schuster, A. 7, 26
second posterior moments 45
Shaw, D. 117
side lobes 26
signal detection
  multiple measurements 167
  one-frequency 21, 90
signal-to-noise *48*
sine transform 7, *16*
Smith, W. H. 162
Sonett, C. P. 148, 151, 157, 158
spectrum absorption 117
stacking brute 163
Student t-distribution 19, 35
  computing 197
  one-frequency 71, 87
  multiple harmonic frequencies 109
  two-frequencies 94
sufficient statistic definition 7, 35, 110
sum rule 10
times discrete *13*
trend elimination 137
Tribus, M. 76
Tukey, J. W. 6, 9, 23, 73
uniform prior 18, 34
units conversion 21, 22
variance 40, 47
Waldmeier, M. 27
weighted averages 190
Welch, P. D. 6
Whittle, P. 110
Wilde, D. J. 50
Wolf's relative sunspot numbers 27, 148
$y(t)$ definition 13
Zellner, A. 55, 73, 76
zero padding 19

GPSR Compliance

The European Union's (EU) General Product Safety Regulation (GPSR) is a set of rules that requires consumer products to be safe and our obligations to ensure this.

If you have any concerns about our products, you can contact us on ProductSafety@springernature.com

In case Publisher is established outside the EU, the EU authorized representative is:

Springer Nature Customer Service Center GmbH
Europaplatz 3
69115 Heidelberg, Germany

Batch number: 10704707

Printed by Printforce, the Netherlands